Optimizing Op Amp Performance

Other Reference Books of Interest by McGraw-Hill

Handbooks

BENSON • *Television Engineering Handbook*
CHEN • *Fuzzy Logic and Neural Network Handbook*
COOMBS • *Printed Circuits Handbook, 4/e*
DI GIACOMO • *Digital Bus Handbook*
DI GIACOMO • *VLSI Handbook*
CHRISTIANSEN • *Electronics Engineers' Handbook, 4/e*
GRAEME • *Photodiode Amplifiers Op Amp Solutions*
HARPER • *Electronic Packaging and Interconnection Handbook, 2/e*
JURAN AND GRYNA • *Juran's Quality Control Handbook*
OSA • *Handbook of Optics, 2/e*
RORABAUGH • *Digital Filter Designer's Handbook, 2/e*
SERGENT AND HARPER • *Hybrid Microelectronic Handbook*
WAYNANT • *Electro-Optics Handbook*
WILLIAMS AND TAYLOR • *Electronic Filter Design Handbook*
ZOMAYA • *Parallel and Distributed Computing Handbook*

Other

ANTOGNETTI AND MASSOBRIO • *Semiconductor Device Modeling with SPICE*
BEST • *Phase-Locked Loops, 3/e*
HECHT • *The Laser Guidebook*
KIELKOWSKI • *Inside SPICE*
SMITH • *Thin-Film Deposition*
SZE • *VLSI Technology*
TSUI • *LSI / VLSI Testability Design*
WOBSCHALL • *Circuit Design for Electronic Instrumentation*
WYATT • *Electro-Optical System Design*

To order or receive additional information on these or any other McGraw-Hill titles, please call 1-800-822-8158 in the United States. In other countries, contact your local McGraw-Hill representative.

WM16XXA

Optimizing Op Amp Performance

Jerald Graeme

McGraw-Hill
New York San Francisco Washington, D.C. Auckland Bogotá
Caracas Lisbon London Madrid Mexico City Milan
Montreal New Delhi San Juan Singapore
Sydney Tokyo Toronto

Library of Congress Cataloging-in-Publication Data

Graeme, Jerald G.
    Optimizing op amp performance / Jerald G. Graeme.
       p.   cm.
    Includes index.
    ISBN 0-07-024522-3
    1. Operational amplifiers. I. Title.
    TK7871.58.06G732   1997
    621.39'5—dc20   96-41207
                                                                                CIP

McGraw-Hill

A Division of The **McGraw·Hill** *Companies*

Copyright © 1997 by The McGraw-Hill Companies, Inc. All rights reserved. Printed in the United States of America. Except as permitted under the United States Copyright Act of 1976, no part of this publication may be reproduced or distributed in any form or by any means, or stored in a data base or retrieval system, without the prior written permission of the publisher.

1 2 3 4 5 6 7 8 9 0   DOC/DOC   9 0 1 0 9 8 7 6

ISBN 0-07-024522-3

The sponsoring editor for this book was Stephen S. Chapman, the editing supervisor was Caroline R. Levine, and the production supervisor was Suzanne W. B. Rapcavage. This book was set in Century Schoolbook by Priscilla Beer of McGraw-Hill's Professional Book Group composition unit. Illustrated by Lola E. Graeme.

Printed and bound by R. R. Donnelley & Sons Company.

McGraw-Hill books are available at special quantity discounts to use as premiums and sales promotions, or for use in corporate training programs. For more information, please write to the Director of Special Sales, McGraw-Hill, 11 West 19th Street, New York, NY 10011. Or contact your local bookstore.

 This book is printed on recycled, acid-free paper containing 10% postconsumer waste.

Information contained in this work has been obtained by The McGraw-Hill Companies, Inc. ("McGraw-Hill") from sources believed to be reliable. However, neither McGraw-Hill nor its authors guarantee the accuracy or completeness of any information published herein, and neither McGraw-Hill nor its authors shall be responsible for any errors, omissions, or damages arising out of use of this information. This work is published with the understanding that McGraw-Hill and its authors are supplying information, but are not attempting to render engineering or other professional services. If such services are required, the assistance of an appropriate professional should be sought.

Contents

Preface   ix

Chapter 1. Performance Analysis, Feedback, and Stability   1

  1.1 Performance Prediction through the Feedback Factor   2
      1.1.1 Noninverting circuit performance and closed-loop gain   2
      1.1.2 The input-referred performance summary   4
  1.2 Feedback Modeling and Bandwidth   6
      1.2.1 Feedback modeling for the noninverting case   6
      1.2.2 Response roll off and loop gain   8
      1.2.3 Bandwidth and the $1/\beta$ intercept   11
      1.2.4 Frequency response of error signals   12
  1.3 Frequency Stability   13
      1.3.1 Frequency stability and the $1/\beta$ intercept   13
      1.3.2 The Bode phase approximation   15
      1.3.3 Intuitive examination of oscillation   18
  1.4 Op Amp Influence on Feedback Networks   21
      1.4.1 Effect of input capacitance   21
      1.4.2 Effect of input inductance   24

Chapter 2. Feedback Modeling and Analysis   29

  2.1 Inverting Configurations   30
      2.1.1 Feedback model for inverting configurations   30
      2.1.2 Feedback analysis of inverting configurations   33
      2.1.3 Noise bandwidth and the inverting configuration   36
  2.2 Positive Feedback Configurations   38
      2.2.1 Positive feedback example   38
      2.2.2 Modeling positive feedback   41
      2.2.3 Positive feedback and the feedback factor   42
  2.3 Dual Input Connections   44
      2.3.1 Dual input example   44
      2.3.2 Modeling dual input connections   46
  2.4 Feedback Analysis Process   48
      2.4.1 Summary of the feedback analysis process   48

	2.4.2 Defining the net feedback factor	50
	2.4.3 Intermediate results	51
	2.4.4 Completing the analysis	52
2.5	Generalized Feedback Model	54
	2.5.1 Generalized model and its performance results	54
	2.5.2 Applying the generalized result	56
2.6	Analysis of Complex Feedback	57
	2.6.1 Multiple amplifiers and the β feedback block	58
	2.6.2 Multiple amplifiers and the A gain block	60
	2.6.3 Variable feedback factors	63
	2.6.4 Multiple input connections	65
	2.6.5 Multiple feedback paths	67

Chapter 3. Power-Supply Bypass — 73

3.1	Power-Supply Bypass Requirement	73
	3.1.1 Noise coupling mechanism	74
	3.1.2 Frequency response of supply noise coupling	76
	3.1.3 Power-supply coupling and frequency stability	78
	3.1.4 Oscillation condition	79
3.2	Selecting the Primary Bypass Capacitor	81
	3.2.1 Fundamental bypass resonance	81
	3.2.2 Graphical evaluation of bypass resonance	83
	3.2.3 Bypass selection	85
3.3	Selecting a Secondary Bypass Capacitor	86
	3.3.1 Bypass capacitor self-resonance	86
	3.3.2 Dual-bypass capacitors	90
	3.3.3 Dual-bypass selection	92
3.4	Bypass Alternatives	94
	3.4.1 Detuning the dual-bypass resonance	94
	3.4.2 Selecting the detuning resistance	97
3.5	Power-Supply Decoupling	100
	3.5.1 Decoupling alternatives	100
	3.5.2 Selecting the decoupling components	104

Chapter 4. Phase Compensation — 107

4.1	Phase Compensation for Capacitance Loading Effects	108
	4.1.1 Capacitance loading and frequency stability	108
	4.1.2 Decoupling the capacitance load	111
	4.1.3 Selecting the decoupling components	113
	4.1.4 Filtering provided by decoupling	115
	4.1.5 Pole-zero compensation for capacitance loading	116
	4.1.6 Selecting the pole-zero compensation components	118
4.2	Phase Compensation for Input Capacitance Effects	120
	4.2.1 Input capacitance and frequency stability	121
	4.2.2 Testing for input capacitance effects	124
	4.2.3 Compensating the inverting configuration	125
	4.2.4 Compensating the noninverting gain peaking	126
	4.2.5 Bandwidth improvement in the noninverting case	128
	4.2.6 Selecting the noninverting compensation	131
	4.2.7 Input capacitance and the differential input configuration	132
	4.2.8 Balancing the differential input configuration	134

4.2.9	Compensating the differential input configuration	135
4.2.10	Alternative noninverting compensation	136
4.3	Multipurpose Phase Compensation	139
4.3.1	Feedback-factor compensation with negative feedback	140
4.3.2	Side effects of feedback-factor compensation	142
4.3.3	Selecting the negative feedback components	144
4.3.4	Feedback-factor compensation for the integrator	145
4.3.5	Positive feedback compensation of voltage followers	148
4.3.6	Selecting the positive feedback components	151
4.3.7	Compensating the differential input configuration	152

Chapter 5. Reducing Radiated Interference   155

5.1	Reducing Electrostatic Coupling	156
5.1.1	Electrostatic shielding	156
5.1.2	Common-mode rejection of electrostatic coupling	156
5.1.3	Common-mode rejection with nondifferential circuits	158
5.2	Reducing Magnetic and RFI Coupling	160
5.2.1	Magnetic shielding	161
5.2.2	Minimizing loop areas	162
5.2.3	Common-mode rejection of magnetic coupling	164
5.3	Reducing Multiple Coupled-Noise Effects	167
5.4	Minimizing Magnetic Field Generation	167

Chapter 6. Distortion and Its Measurement   171

6.1	The Nature of Op Amp Distortion	172
6.1.1	Op amp distortion signal	173
6.1.2	Input-referred op amp distortion	175
6.2	Basic Distortion Measurement	177
6.2.1	Spectrum analyzer measurement	178
6.2.2	Distortion analyzer measurement	179
6.2.3	Basic measurement limitations	181
6.3	Feedback Separation of Distortion Signals	181
6.3.1	Signal separation and the voltage follower	182
6.3.2	Signal separation and noninverting configurations	183
6.3.3	Signal separation and inverting configurations	185
6.4	Direct Measurement of Feedback Error Signal	187
6.4.1	Direct measurement and inverting configurations	188
6.4.2	Direct measurement and the voltage follower	190
6.4.3	Direct measurement and noninverting configurations	192
6.5	Selective Amplification of Feedback Error Signal	194
6.5.1	Selective amplification and the voltage follower	195
6.5.2	Bandwidth of selective amplification	197
6.5.3	Selective amplification and noninverting configurations	199
6.5.4	Selective amplification and inverting configurations	201
6.6	Selective Amplification Alternatives	202
6.6.1	Combined signal separation and selective amplification	203
6.6.2	Variable selective amplification	205
6.7	Bootstrap Isolation of Feedback Error Signal	207
6.7.1	Bootstrap isolation and the voltage follower	208
6.7.2	Distortion equivalence of the bootstrap measurement	211

6.7.3	Bootstrap isolation and the noninverting configuration	212
6.7.4	Distortion equivalence and the noninverting case	214
6.7.5	Combined bootstrap isolation and selective amplification	215

Glossary 219
Index 223

Preface

The op amp serves as a universal gain block to greatly simplify analog design. Just adding a few feedback elements generally establishes a near ideal closed-loop gain or frequency response without the need for extensive circuit analysis. The op amp's very high open-loop gain and restricted error sources make this design process possible and permit a reliance upon very basic design equations. However all too often, this simplistic design approach suboptimizes the end result or the ideal condition fails, resulting in unanticipated errors, bandwidth limitations, oscillation, noise, and/or distortion. This book addresses those conditions and develops simple design equations and models that extend op amp insight to a second level of detail. That level permits design optimization, again without extensive analysis.

Fundamentally, the op amp serves as an ideal gain block awaiting feedback direction of its input-to-output response. More than any other design factor, this feedback also dictates the op amp's general performance and frequency stability conditions. The resulting feedback factor, β, defines the $1/\beta$ gain that amplifies the numerous input-referred errors of the op amp. Indirectly, this gain also defines the closed-loop bandwidth of an application circuit. Graphical analysis quantifies this bandwidth through the frequency response plots representing the available open-loop gain and the feedback demand for that gain. This same analysis indicates frequency stability characteristics through the slopes of the two plots and the corresponding feedback phase shift that they predict. Design equations reduce the graphical analysis results to a simple form for universal application to all op amp circuits.

However, the application of these results first requires the determination a given circuit's feedback factor. Feedback modeling defines β and dramatically simplifies op amp circuit analysis. Given the corresponding model, a circuit's performance analysis reduces to the determination of voltage divider ratios and these ratios adapt standardized performance results to any specific circuit considered. For the noninverting op amp configuration, the classic Black feedback model applies directly. Other configurations require extensions of this model to represent their feedback behavior, as demonstrated by a series of examples. These examples adapt the Black model to inverting circuits, positive feedback connections, and dual input connections. In this process, a generalized modeling approach evolves that permits rapid analysis of any op amp configuration. Also, standardized performance results develop that immediately apply to any modeled circuit. A generalized feedback model summarizes the modeling process and the standardized results. This model directly represents most op amp circuits and further extensions adapt that model to the most complex circuits. Such extensions adapt to complex feedback connections that include multiple gain elements, gain in the feedback path, variable feedback, and multiple feedback connections.

As mentioned, the powerful feedback analysis described above predicts the frequency-stability conditions of an op amp circuit, but only those established by the amplifier's primary feedback loop. When oscillation or other degraded stability conditions occur, one should first consider the parasitic feedback loop developed through the power supply connections. Often, greater attention to power supply bypass corrects these conditions at the outset and does so with far less effort. Habit rather than analysis often erroneously guides the mundane selection of this bypass and can result in unexpected stability problems. Analysis of the parasitic, supply-line feedback reveals a complex relationship between an op amp and its supply line impedances. There, inductive effects and multiple impedance resonances compromise the simple capacitance shunting desired from the bypass. Resulting power line signals couple through the op amp to its input via the amplifier's finite power-supply rejection. To minimize this parasitic feedback, analyzing the bypass effects yields simple design equations that optimize the bypass selection. Discussions of dual bypass capacitors, bypass detuning, and power supply decoupling expand the bypass alternatives.

After assuring good bypass conditions, consider phase compensation for the correction of any remaining stability problems. The greatly varied applications of op amps routinely encounter response ringing or oscillation, and in such cases, the general-purpose phase compensation included within the op amp often proves to be insufficient for the application. Then, external phase compensation must be added. However, most op amps lack the provision for external adjustment of their internal compensation. Then, an external adjustment of the amplifier's feedback network provides the required compensation instead. This book describes varied feedback compensation methods that modify either the effective open-loop response of the circuit or the circuit's 1/β response. In each case, the phase compensation selection focuses upon the control of the feedback phase shift occurring at the 1/β intercept with A_{OL}.

As described before, an op amp circuit's 1/β response accurately predicts a given application's output errors resulting from the op amp's input-referred error sources. However, much like the stability issue before, factors outside the amplifier's primary feedback loop also affect output noise. The 1/β response does amplify the op amp's input noise voltage and generally produces the major component of output noise. Still, other less obvious noise sources also contribute to that output noise due to coupling from external noise sources. Then, diminishing returns eventually limit the noise reduction achieved through measures focused upon the visible, hard-wired portion of a circuit. As a companion part of every circuit, an invisible, but very real, circuit portion couples radiated noise into the signal path. There, electrostatic and magnetic coupling impose a background noise floor that requires attention to the amplifier's environment instead of the intended circuit. Physical separation and shielding remain the first defense against this coupling but op amp circuits present additional noise reduction opportunities. Differential inputs, impedance balancing, and loop minimization all aid in the reduction of this coupled noise. Evaluations of the coupling mechanisms guide the application of these techniques.

The feedback analyses previously described include the effects of the most common op amp error sources but do not directly address distortion. However, once defined, the distortion error also resolves in the 1/β analysis through the amplifier's input error signal. Many characteristics of an op amp circuit introduce distortion, but feedback consolidates their effects in that error signal and does so in the background of only a small fraction of the test signal. This consolidation makes op amp distortion an input-referred characteristic, and two conveniences result from this input-referred result. First, the input-referred distortion measured in a given application configuration permits prediction of the output-referred distortion for other configurations. This convenience parallels that of an op amp's input-offset voltage where the measurement of the input-referred characteristic permits prediction of the output-referred offset for any circuit configuration.

In the second convenience, the feedback separation of the op amp's distortion component dramatically reduces test equipment requirements. There, the input error signal containing the distortion components separates them from the circuit's large output signal. This signal separation largely removes the distortion introduced by an input signal generator and dramatically reduces the dynamic range required of the distortion analyzer. Three fundamental measurement techniques exploit this signal separation. Direct measurement, selective amplification, and bootstrap isolation all focus upon the separated input error signal to improve the resolution of distortion measurements. In the simplest case, a direct measurement of the feedback error signal yields the amplifier's input-referred distortion but this approach requires an added instrumentation amplifier for noninverting configurations. In the next case, the amplifier-under-test selectively amplifies the input error signal to increase the relative significance of the distortion in the amplifier's output signal. With the final technique, bootstrap isolation makes the feedback error signal ground-referenced for noninverting configurations, avoiding the common-mode signal that would otherwise complicate the measurement.

I wish to thank the many op amp users with whom I have discussed application requirements over the years. Their inquiries led to the investigation and development of much of the material presented here. My thanks to *EDN* and *Electronic Design* for publishing the articles that initiated many of these inquiries. Finally, I wish to thank my wife Lola for her accurate and attractive rendering of the illustrations and for the rewarding feeling of mutual involvement in preparing this book.

Jerald Graeme

Chapter

1

Performance Analysis, Feedback, and Stability

The chapter title suggests an ambitious goal to encompass performance analysis, feedback, and stability in one treatment. Yet this combination reflects the analytical power gained through knowledge of an op amp circuit's feedback factor. Feedback dictates the performance of an op amp, both in function and in quality. The major specifications of the amplifier describe an open-loop device awaiting feedback direction of the end circuit's function. Just how well the amplifier performs the function reflects through the feedback interaction with the open-loop error specifications. Fortunately, most open-loop errors simply reflect to the circuit output amplified by the reciprocal of the circuit's feedback factor.

Amplifier bandwidth limits this simple relationship, but the feedback factor defines this limit as well. Above a certain frequency, the amplifier lacks sufficient gain to continue amplification of signal and errors alike. Graphical analysis defines this frequency limit through plots representing available amplifier gain and the feedback demand for that gain. This same analysis indicates frequency stability characteristics for op amp circuits. There just the slopes of the plots indicate the phase shift in the feedback loop. Thus the feedback factor of an op amp circuit is a powerful performance indicator.

The determination of a circuit's feedback factor depends upon feedback modeling, and for the noninverting op amp configuration the basic feedback model applies directly. Using this configuration, this chapter demonstrates the performance, feedback, and stability concepts common to all op amp configurations. Later Chap. 2 extends feedback modeling to the general op amp case and also extends the concepts and conclusions of this chapter to all other op amp configurations.

1.1 Performance Prediction through the Feedback Factor

More than any other parameter, the feedback factor of an op amp application defines the circuit performance.[1] The feedback factor first sets the gain received by the input-referred errors of the amplifier. These open-loop errors include offset voltage, noise, and the error signals generated by finite open-loop gain, common-mode rejection, and power-supply rejection.

1.1.1 Noninverting circuit performance and closed-loop gain

For the noninverting op amp configuration, a convenient relationship between closed-loop gain and feedback factor simplifies performance analysis. There the gain of the application circuit itself sets the amplification of input-referred errors and determines the circuit bandwidth. Shown in Fig. 1.1 as a voltage amplifier, this noninverting circuit produces the familiar, ideal closed-loop gain of $A_{CLi} = (R_1 + R_2)/R_1$. This gain amplifies both the input signal e_i and the differential input error e_{id} of the op amp. Simply multiplying e_{id} by A_{CLi} defines the resulting output error. Later examination adds frequency dependence to this simple relationship.

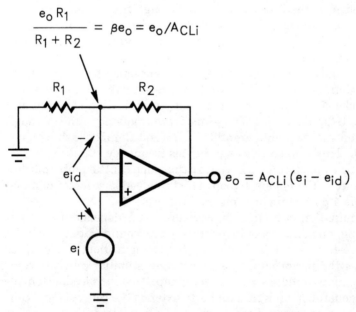

Figure 1.1 Noninverting op amp connections amplify input signal e_i and error signal e_{id} by a gain of $A_{CLi} = 1/\beta$.

Performance Analysis, Feedback, and Stability

The fundamental mechanism relating input and output errors lies in the feedback factor. That factor is the fraction of the amplifier output signal fed back to the amplifier input. In the figure, a feedback voltage divider defines this fraction through the output-to-input transfer response

$$\beta e_o = \frac{e_o R_1}{R_1 + R_2}$$

This defines ß as simply the voltage divider ratio of the feedback network, $\beta = R_1/(R_1 + R_2)$. Comparing this result with $A_{CLi} = (R_1 + R_2)/R_1$ shows that $A_{CLi} = 1/\beta$ for the noninverting case.

However, general error analysis depends on ß rather than an A_{CLi}, as emphasized with the model of Fig. 1.2. This model represents the noninverting op amp connection by an amplifier with input error signal e_{id} and with feedback transmission factor ß. There ß determines the signal βe_o fed back to the amplifier input from the output signal e_o. Writing a loop equation for the model shows that

$$e_o = (1/\beta)(e_i - e_{id})$$

In this result, a gain of 1/ß amplifies both e_i and e_{id}. Thus the circuit of Fig. 1.1 and the model of Fig. 1.2 agree for purposes of input-to-output transmission of amplifier signals.

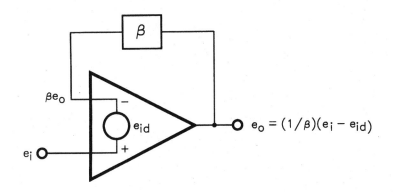

$$e_{id} = V_{OS} + I_{B+}R_{S+} - I_{B-}R_{S-} + e_n + \frac{e_o}{A} + \frac{e_{icm}}{CMRR} + \frac{\delta V_S}{PSRR}$$

Figure 1.2 Input error e_{id} amplified by 1/ß includes effects of major performance characteristics of an op amp.

1.1.2 The input-referred performance summary

The simple 1/ß relationship between input and output errors predicts the output errors resulting from almost all amplifier performance characteristics. Each of these characteristics produces an input-referred error source for the op amp as combined in

$$e_{id} = V_{OS} + I_{B+}R_{S+} + I_{B-}R_{S-} + e_n + e_o/A + e_{icm}/\text{CMRR} + \delta V_S/\text{PSRR}$$

Error terms included in this equation cover the effects of the op amp input offset voltage, input bias currents, input noise voltage, open-loop gain, common-mode rejection, and power-supply rejection. There the R_{S+} and R_{S-} resistances represent the source resistances presented to the op amp's two inputs. The last three error terms include circuit signals, which are the output voltage, the common-mode voltage, and the power-supply voltage change.

For each error term the definitions of op amp performance characteristics define the input-referred source either directly or in combination with simple analyses. Definitions directly classify V_{OS}, I_{B+}, I_{B-}, and e_n as input-referred error sources. Next, open-loop gain is defined as the ratio of the output voltage to the differential input voltage that produces it, $A = e_o/\delta e_{id}$. This makes the corresponding input-referred error signal $\delta e_{id} = e_o/A$. Similarly, the power-supply rejection ratio equals the ratio of a power supply change to the resulting change in differential input voltage. Thus $\text{PSRR} = \delta V_S/\delta e_{id}$ and the associated input-referred error is $\delta e_{id} = \delta V_S/\text{PSRR}$.

For the common-mode rejection ratio, the relationship between definition and input error requires closer examination. The common-mode rejection ratio is defined as the ratio of the differential gain to the common-mode gain A_D/A_{CM}. For an op amp, the differential gain is simply the open-loop gain A. Then, $\text{CMRR} = A/A_{CM}$, and rewriting this shows the common-mode gain to be $A_{CM} = A/\text{CMRR}$. However, by definition $A_{CM} = e_{ocm}/e_{icm}$, where e_{ocm} is the output signal resulting from an input common-mode signal e_{icm}. Combining the two A_{CM} equations results in $e_{ocm} = Ae_{icm}/\text{CMRR}$. To support this component of output voltage, the op amp develops another gain error signal in e_{id}. As before, the resulting e_{id} component equals the associated output voltage divided by the open-loop gain. Dividing the preceding e_{ocm} expression by open-loop gain A defines the input-referred CMRR error as e_{icm}/CMRR.

Closer examination also clarifies the source resistances R_{S+} and R_{S-} of the e_{id} equation. In the simplest case, a source resistance is just the output resistance of a signal source that drives a circuit

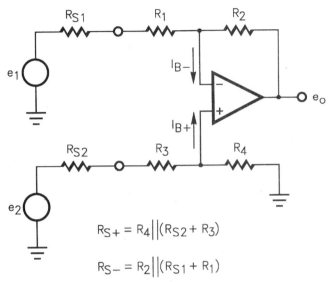

Figure 1.3 Scaling and feedback resistors alter source resistances presented to the input bias currents of an op amp.

input. However, for op amp circuits, scaling and feedback resistors alter the net resistances presented to the amplifier's inputs. The difference amplifier connection well illustrates this, as shown in Fig. 1.3. There scaling resistors R_3 and R_4 alter the resistance presented to the amplifier's noninverting input, and feedback resistors R_1 and R_2 alter that presented to the inverting input. Signal sources e_1 and e_2 drive the difference amplifier's inputs through conventional source resistances R_{S_1} and R_{S_2}. Together, these scaling, feedback, and source resistances determine the net resistances presented to the op amp's input currents I_{B+} and I_{B-}. Current I_{B+} divides between two paths to ground through R_4 and the $R_3 + R_{S_2}$ combination. Here the $R_3 + R_{S_2}$ path returns to ground through the zero resistance of the e_2 ideal source. Thus for the e_{id} equation, $R_{S+} = R_4 ||(R_3 + R_{S_2})$. Analogously, I_{B-} divides between the path through R_2 and that through $R_1 + R_{S_1}$. In this case, R_2 departs from the analogy by returning to the op amp output instead of to ground. However, the low output impedance of the op amp produces an equivalent result for this resistance evaluation and $R_{S-} = R_2 ||(R_1 + R_{S_1})$.

Together, the error terms of the model in Fig. 1.2 provide a fairly complete representation of op amp performance limits. However, the e_{id} expression does not specifically list errors due to distortion, bandwidth, and slew-rate limiting. As described in Chap. 6, e_{id} actually includes the amplifier's distortion error in the gain and CMRR error signals. Still, a circuit's bandwidth limit restricts the effects of the e_{id} error sources at higher frequencies and slew-rate limiting imposes a secondary bandwidth limit for large signal operation. Feedback factor analysis treats these bandwidth effects in the next section.

Up to the circuit's bandwidth limit, each input-referred error term of the model in Fig. 1.2 reflects to the amplifier output through a gain equal to 1/ß. Multiplication of the error terms by 1/ß produces some familiar results. The output error due to the finite open-loop gain becomes $e_o/Aß$. This shows that the gain error reduces the output e_o by a fraction of that output, and this fraction equals the reciprocal of the loop gain $Aß$. As will be described, the decline of A with frequency makes this error rise, and this shapes the closed-loop frequency response of the circuit. Similar multiplication of the input noise error defines the output noise as $e_n/ß$, leading to the term "noise gain" for 1/ß. However, this description of 1/ß only holds under the bandwidth limits to be described. For both the loop gain and the noise errors, greater visibility results through the frequency response analysis presented hereafter.

1.2 Feedback Modeling and Bandwidth

Section 1.1 presents the 1/ß relationship between input-referred op amp error sources and the resulting output errors. However, the frequency dependence of amplifier gain modifies this simple, initial relationship. Amplifier response roll off defines a bandwidth limit for both signal and error sources. This reduces the output error effects of all error sources except for the dc errors of V_{OS}, $I_{B+}R_{S+}$, and $I_{B-}R_{S-}$. Amplifier gain, noise, CMRR, and PSRR produce ac errors, and their output effects depend upon the circuit's frequency response. More complete feedback modeling defines this frequency response through the noninverting amplifier example of this chapter. The response results developed in this section extend to any op amp configuration through a standardized response denominator as developed in Chap. 2.

1.2.1 Feedback modeling for the noninverting case

Figure 1.4 presents the generalized noninverting connection with the feedback network shown as the generalized Z_1 and Z_2 rather than the previous resistors. Also, redrawing the amplifier configuration as shown highlights the voltage divider action of the feedback network. The network's divider action again displays the fraction of the amplifier output fed back to the amplifier input. In preparation for the next modeling step, the figure reduces the amplifier input error signal e_{id}

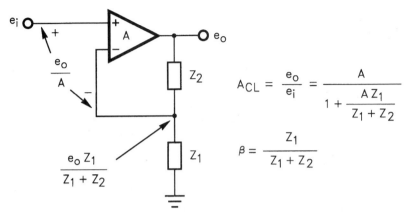

Figure 1.4 Redrawing the op amp circuit and reducing e_{id} to gain error signal e_o/A prepares the circuit for feedback modeling.

to just the open-loop gain error e_o/A. Feedback modeling focuses upon just the gain error and its related frequency characteristics. Still, analysis of this one error signal suffices to define the frequency response for use with the previous multierror analysis.

Loop analysis of Fig. 1.4 defines the noninverting circuit's transfer response as

$$A_{\text{CL}} = \frac{e_o}{e_i} = \frac{A}{1 + AZ_1/(Z_1 + Z_2)}$$

Gain A in this expression contains the frequency dependence that shapes the circuit's frequency response. Note that the denominator of this response contains the feedback factor $Z_1/(Z_1 + Z_2)$. This makes the denominator $1 + A\beta$, and this denominator relates the circuit to the feedback model presented next.

To more completely analyze the noninverting circuit, Fig. 1.5 replaces the op amp of Fig. 1.4 with a gain block and a summation element. Also, a feedback block replaces the feedback network from before. The gain block represents the amplifier open-loop gain and the summation models the differential action of the op amp inputs. Op amp open-loop gain amplifies the differential signal between the two amplifier inputs and opposite polarities as the model's summation inputs reproduce the differential action. For these model inputs, the polarity assignments match the polarities of the corresponding op amp inputs. With these assignments, the summation extracts the differential signal through subtraction. The model then supplies the differential signal to the gain block, and this block drives the feedback block ß. For op amps this classic feedback model, initially devel-

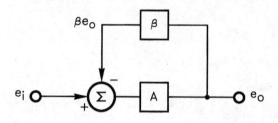

$$A_{CL} = \frac{e_o}{e_i} = \frac{A}{1 + A\beta} = \frac{1/\beta}{1 + 1/A\beta}$$

$$A_{CL\,i} = 1/\beta, \text{ for } A \longrightarrow \infty$$

Figure 1.5 Black's classic feedback model reproduces the A_{CL} transfer response of noninverting op amp configurations.

oped by Black,[2] only represents the noninverting case. However, modifications to the model adapt it to other configurations in the next chapter.

Comparison of the circuit and model responses demonstrates the model's validity. The model amplifies the difference between the summation inputs by gain A to produce the output signal. This results in $e_o = A(e_i - \beta e_o)$, and solving for e_o/e_i defines the modeled transfer response as

$$A_{CL} = \frac{e_o}{e_i} = \frac{A}{1 + A\beta}$$

Comparison of terms in the A_{CL} equations for the model described and the circuit shown in Fig. 1.4 again shows the feedback factor to be $\beta = Z_1/(Z_1 + Z_2)$, validating the model.

1.2.2 Response roll off and loop gain

Further analysis of the A_{CL} result defines the op amp frequency response and stability conditions.[3] This added performance information depends upon the denominator of the A_{CL} response and not upon the specific noninverting case considered here. Conclusions based upon this denominator extend to all other op amp configurations as described in Chap. 2. Rewriting the A_{CL} equation yields

$$A_{CL} = \frac{1/\beta}{1 + 1/A\beta}$$

Then the response numerator expresses the ideal closed-loop gain of the noninverting configuration $A_{CLi} = 1/\beta$, and the denominator expresses the gain's frequency dependence through A and β.

Other op amp configurations produce different A_{CLi} numerators, but always the same $1 + 1/A\beta$ denominator. This common denominator unifies bandwidth and stability characteristics for all op amp configurations through feedback factor β. As demonstrated in the next chapter, all op amp configurations produce a closed-loop response of

$$A_{CL} = \frac{A_{CLi}}{1 + 1/A\beta}$$

Writing a given configuration's response with a denominator in this form immediately identifies the ideal response A_{CLi} as the numerator. It also links the configuration directly to the denominator-based bandwidth results and stability criteria that follow.

The frequency dependencies of A and β combine to set a configuration's frequency response. At low frequencies, the high level of open-loop gain A reduces the denominator to $1 + 1/A\beta \sim 1$. Then the circuit response simplifies to the ideal gain of A_{CLi}. At higher frequencies, the op amp open-loop gain drops, causing this denominator to increase, and A_{CL} declines from its ideal value of A_{CLi}. Similarly, a high-frequency drop in β would add to the A_{CL} decline, but, initially, assuming a constant β simplifies the analysis. Constant β results with the resistive feedback networks common to the majority of applications. Reactive rather than resistive feedback slightly modifies the bandwidth conclusions developed here and later examples describe this effect. However, such reactive feedback does not alter the frequency stability conditions developed through this resistive feedback example.

For the resistive feedback case, the open-loop gain decline with frequency produces the circuit's bandwidth limit, as illustrated in Fig. 1.6. The plot displays the frequency responses of all three variables in the A_{CL} equation. Shown are the closed-loop gain A_{CL}, the open-loop gain A, and $1/\beta$ as a function of frequency. The graphical interaction of these variables provides visual insight into bandwidth and frequency stability limits. The heavier curve represents the resulting closed-loop response A_{CL} and displays this gain's high-frequency roll off. The loop gain of the circuit, $A\beta$ in the preceding A_{CL} denominator, represents the amplifier gain resource available to maintain the ideal closed-loop response. In Fig. 1.6 the shaded area of the graph high-

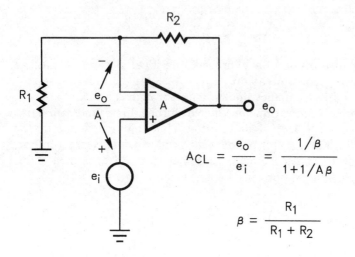

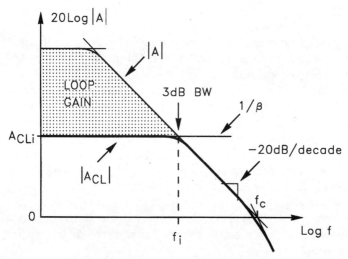

Figure 1.6 Graphical analysis with A and $1/\beta$ curves of a circuit defines closed-loop bandwidth.

lights this gain. At any given frequency, the corresponding loop gain equals the vertical distance separating the A and $1/\beta$ curves. The logarithmic scale of the graph makes this distance $\log A - \log(1/\beta) = \log(A\beta)$. Loop gain $A\beta$ represents the amplifier's reserve capacity to supply the feedback demand for gain, and where loop gain drops below unity, the A_{CL} curve shown drops from the ideal A_{CLi} level.

The A and $1/ß$ curves display this loop gain limit graphically. Here the $1/ß$ curve represents the feedback demand for open-loop gain, and the loop gain meets this demand as long as the $1/ß$ curve remains below the open-loop response curve. However, at higher frequencies, the open-loop gain curve falls below the $1/ß$ level. There the feedback demand exceeds the available amplifier gain, making A_{CL} roll off, following the amplifier open-loop response. This response roll off follows a -20-dB per decade slope for the single-pole response characteristic of typical op amps.

1.2.3 Bandwidth and the 1/ß intercept

The bandwidth limit of most op amp circuits occurs at the $1/ß$ intercept with the open-loop gain curve. Some circuits reduce bandwidth further, through reactive feedback elements, but all op amp circuits encounter a bandwidth limit at the $1/ß$ intercept. Figure 1.6 illustrates this intercept and the coincident roll off of the A_{CL} response. By definition, the 3-dB bandwidth limit occurs where A_{CL} drops from its dc value to 0.707 times that value. Analysis shows that this condition results at the intersection of the A and $1/ß$ curves. These curves are actually magnitude responses, and at their intersection, their magnitudes are the same, or $|A| = |1/ß|$. Rearranging this result shows that the intercept at f_i occurs where the loop gain is $|Aß| = 1$. A phase shift of $-90°$ accompanies this unity-gain magnitude because of the single-pole roll off of gain A. Then $Aß = -j1$ at the intercept, and the denominator of the A_{CL} equation becomes $1 + 1/Aß = 1 + j1$. Then

$$A_{CL} = \frac{A_{CLi}}{1+j1}, \quad \text{at } f_i$$

The $\sqrt{2}$ magnitude of the preceding denominator drops the circuit gain magnitude from A_{CLi} to $0.707A_{CLi}$. Thus for frequency-independent feedback factors, the 3-dB bandwidth occurs at the intercept frequency f_i. With frequency-dependent feedback factors, the closed-loop response still rolls off following the intercept, but this point may not be the 3-dB bandwidth limit. In those cases, peaking or additional roll off in the closed-loop response curve moves the actual 3-dB point away from f_i.

For the more common op amp applications, constant feedback factors permit a simple equation for the 3-dB bandwidth. Single-pole responses characterize the open-loop roll offs of most op amps, and virtually all $1/ß$ intercepts occur in this single-pole range. There the single pole makes the gain magnitude simply $|A| = f_c/f$, where f_c is the unity-gain crossover frequency of the amplifier. Then at the intercept, where $|A| = 1/ß$ and $f = f_i$, the last equation becomes $1/ß = f_c/f_i$. Solving for f_i defines the 3-dB bandwidth for most op amp applications as

$$\text{BW} = f_i = ß f_c$$

This result holds for all op amp applications having frequency-independent ß and single-pole op amp roll off.

Technically this bandwidth limit portrays only the small-signal performance of an op amp. In large-signal applications, slew-rate limiting often sets a lesser bandwidth limit, especially in lower-gain, large-signal applications. There the slew-rate limit S_r imposes a power bandwidth limit of $\text{BW}_P = S_r/2\pi E_{op}$, where E_{op} is the peak value of the output voltage swing. This limit represents the only major performance characteristic of an op amp not directly related to the feedback factor ß. However, an indirect relationship still links large-signal bandwidth and ß. The value of ß helps determine which bandwidth limit, BW or BW_P, applies in a given application. Both bandwidth limits set performance boundaries, and the lower of the two prevails in large-signal applications. Higher values of ß imply lower closed-loop gains and increase the frequency boundary set by $\text{BW} = ß f_c$. Then, for large signals, $\text{BW}_P = S_r/2\pi E_{op}$ generally produces the lower of the two boundaries and sets the circuit bandwidth. Conversely, lower values of ß reduce the $\text{BW} = ß f_c$ boundary, making this the dominant limit. For a given application, compare the two limits to determine which applies.

1.2.4 Frequency response of error signals

The frequency dependence defined by the 1/ß intercept also applies to the ac error sources of the analysis in Sec. 1.1.2. That analysis showed that the input-referred errors of op amps transfer to the amplifier output through a gain of 1/ß. However, 1/ß does not include the high-frequency limitations of the amplifier. Thus the earlier analysis remains valid only for frequencies up to the 1/ß intercept at f_i. Above this frequency, the amplifier lacks sufficient gain to amplify input error sources by the 1/ß gain that applies to lower frequencies. The bandwidth limit $\text{BW} = ß f_c$ marks a response roll off that reduces amplification of signal and error alike. Here the limited magnitudes of error signals always invoke the small-signal, rather than slew-rate, bandwidth limit.

This error signal roll off produces the previously mentioned distinction between 1/ß and "noise gain." Beyond the intercept, the gain supplied to noise follows the amplifier response roll off even though the 1/ß curve continues uninterrupted. For A_{CL} the response roll off results from the denominator of this gain's equation. For error signal gain, adding this denominator to the original 1/ß gain inserts the frequency dependence. This makes the closed-loop error gain

$$A_{CLe} = \frac{1/ß}{1 + 1/Aß}$$

Here the added frequency dependence reduces the higher-frequency output errors calculated for the noise, CMRR, and PSRR error sources of Sec. 1.1.2.

For the noninverting case considered here, $A_{CLe} = A_{CL}$, but for other cases A_{CL} varies. The error gain A_{CLe}, however, remains the same for all cases. This gain always equals 1/ß up to this curve's intercept with the amplifier open-loop response and, then, rolls off with that response. Note that A_{CLe} depends only upon the variables ß and A. Any feedback model with ß and A blocks configured like Fig. 1.5 yields the same expression for A_{CLe}. Examples in the next chapter demonstrate this fact.

1.3 Frequency Stability

The ac performance indications of the feedback factor also predict op amp frequency stability. For this added purpose, the response plots that define bandwidth also communicate the phase shift of the feedback loop. Excess phase shift promotes response ringing or oscillation, and the plot slopes indicate this phase shift directly. Mathematical analysis defines the stability indicators applied to the plots, and an intuitive evaluation verifies these indicators.

1.3.1 Frequency stability and the 1/ß intercept

Response plots like that of Fig. 1.6 permit frequency stability evaluation directly from the curve slopes. Specifically, the slopes of the A and 1/ß curves at the intercept indicate the phase shift for a critical feedback condition. As mentioned in Sec. 1.2.3, the intercept corresponds to a loop gain magnitude of $|Aß| = 1$. If the accompanying loop phase shift reaches 180°, this adds a minus sign to the actual loop gain, making $Aß = -1$. Then the denominator of the A_{CL} equation reduces to $1 + 1/Aß = 0$, making $A_{CL} = A_{CLi}/(1 + 1/Aß)$ infinite. With infinite gain, a circuit supports an output signal in the absence of an input signal. In other words, the circuit oscillates, and it does so at the intercept frequency f_i.

The relative slopes of the gain magnitude and 1/ß curves reflect the phase shift of the feedback loop. There the relationship between response slope and phase shift follows from the basic effects of response poles and zeros. A pole creates a −20-dB per decade response slope and −90° of phase shift, and a zero produces the same effects with opposite polarities. Additional poles and zeros simply increase response slope and phase shift in increments of the same magnitudes. The slope and phase correlation accurately predicts the loop phase shift when the critical intercept remains well separated from response break frequencies. However, any break frequency of the amplifier or feedback network residing within a frequency decade of the intercept requires a more detailed analysis, as described later. Even in those cases, the response slopes provide an initial insight into probable stability behavior.

Relying upon the slope and phase correlation, the rate-of-closure guideline quickly approximates the phase shift of Aß. The rate of closure is simply the difference between the slopes of the A and 1/ß curves at their intercept. Both slopes communicate phase shift, and the slope difference indicates the net phase shift of the feedback loop. Figure 1.7 illustrates the slope and phase correspondence for two common feedback cases. There two 1/ß curves having different slopes intercept the gain magnitude curve $|A|$. The $1/ß_1$ curve has the zero slope of resistive feedback networks and the rate of closure depends only upon the gain magnitude curve. This magnitude curve follows the −20-dB per decade slope common to most op amp responses. Together the two curves develop a 20-dB per decade slope difference, or rate of closure, for 90° of Aß phase shift. In the feedback loop, the phase inversion of the op amp adds another 180° for a net phase shift of 270°. This leaves a phase margin of $\phi_m = 90°$ from the 360° needed to support oscillation. For op amp stability analysis, the 180° phase shift from the amplifier phase inversion is automatic. Thus op amp phase analysis simplifies, replacing the normal 360° stability criterion with a criterion of 180° of feedback phase shift. This simplified convention applies in the examples that follow.

The second 1/ß curve of Fig. 1.7 reflects the feedback condition of the basic differentiator circuit which produces a feedback demand represented in the figure by the $1/ß_2$ curve. That curve slopes upward at +20 dB per decade and intercepts the $|A|$ curve where the slope difference is 40 dB per decade. Then the rate-of-closure guideline indicates a feedback phase shift of 180°, leaving zero phase margin. This confirms the inherent oscillation of the undegenerated differentiator circuit.

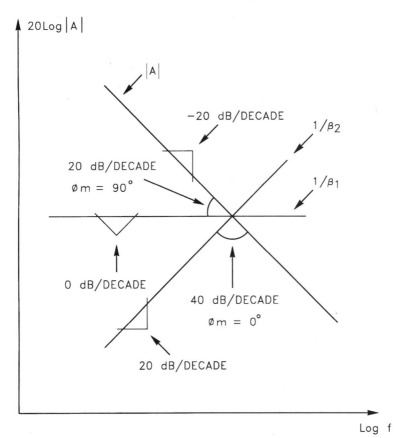

Figure 1.7 Plotted together, 1/ß and open-loop gain curves display circuit frequency stability conditions through the curves' rate of closure.

1.3.2 The Bode phase approximation

As mentioned, the rate-of-closure criterion predicts the Aß phase shift at the intercept accurately when no response break frequencies occur within a decade of the intercept. Other cases require more detailed phase analysis but the Bode phase approximation[4] retains analytical simplicity for such cases. Shown in Fig. 1.8, this approximation predicts the phase effect of a response singularity through a straight-line approximation that deviates from the actual phase by no more than 5.7°.

The actual phase shift introduced by the pole illustrated at f_p progresses through the arctangent curve shown in the figure and expressed by $\phi_p = \arctan(f/f_p)$. However, the Bode approximation provides quicker, visual feedback when examining response plots. This approximation follows a straight line having a slope of $-45°$ per

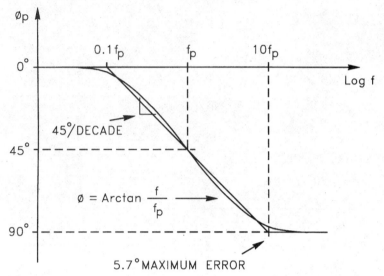

Figure 1.8 Bode phase approximation refines phase analysis for cases where rate-of-closure criterion loses accuracy.

decade and centered on the frequency f_p, where the phase shift is 45°. From there the approximation line predicts 0° at $0.1f_p$ and the full 90° at $10f_p$. Just these three reference points provide a quick visual indication of the phase effect that a given response break produces at a frequency of interest. Outside the band of $0.1f_p$ to $10f_p$, a response break produces little influence. Within this band, visual extrapolation approximates the phase shift. For example, consider a point midway between the f_p and $10f_p$ marks of the log f scale. Note that this midpoint is a linear measure on the log scale. However, application of the phase approximation requires no logarithmic conversion and visual perception of distance applies directly. At this midpoint, the phase approximation indicates a phase shift of 45° + 0.5(45°) = 67.5°. Similarly, at a point two-tenths of the way between $0.1f_p$ and f_p the approximation indicates 0.2(45°) = 9°. These analyses require no knowledge of the actual frequencies represented by the example points. In contrast, exact analysis with the arctangent relationship first requires conversion of the linear distance observed into the equivalent frequency of the log f scale. Then the arctangent expression must be evaluated.

Figure 1.9 illustrates the application of the Bode phase approximation to the stability indication of the 1/ß intercept. In the figure the intercept occurs where the open-loop gain response has a slope of −20 dB per decade. The rate-of-closure guideline suggests 90° of loop phase shift. However, a second amplifier pole at f_p develops a −40-dB

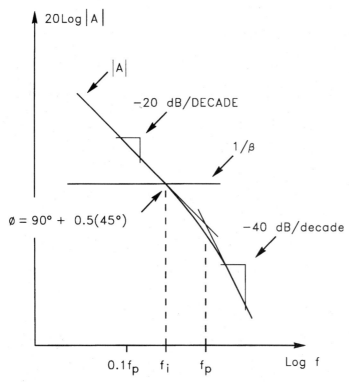

Figure 1.9 Application of the Bode approximation defines phase effects of response breaks that occur less than a decade from the intercept at f_i.

per decade slope later in the open-loop gain response. In this example, the pole at f_p occurs less than a decade in frequency from the intercept, so this limited separation compromises the simple rate-of-closure indication. Then use of the Bode phase approximation refines the estimate of the phase effect of f_p at the intercept f_i. As shown, f_p occurs above f_i, so the approximation first indicates that the effect at f_i is less than 45°. Next a refinement of this initial estimate follows from the linear distance separating f_p and f_i on the plot. This linear distance represents a fraction of a frequency decade, and that fraction equals this distance divided by that separating f_p and $0.1f_p$. Visual reading of the example shown places f_i about midway between f_p and $0.1f_p$. This communicates a phase effect from f_p of $0.5(45°) = 22.5°$ at the intercept frequency f_i. Adding this to the 90° produced by the -20-dB per decade slope of $|A|$ results in a net loop phase shift of 112.5°. This leaves 67.5° of phase margin from the 180° of feedback phase shift required for oscillation.

1.3.3 Intuitive examination of oscillation

With op amps, conventional insight into the cause of amplifier oscillation can be misleading. In the general amplifier case, high gain combined with high phase shift promotes oscillation. However, in the op amp case, these conditions often coexist without producing instability. The distinction lies in the simultaneous gain and phase conditions required for op amp oscillation. At lower frequencies, high loop gain prevents oscillation by attenuating the amplifier's input error signal. At higher frequencies, a lack of loop gain restricts the output signal to similarly prevent oscillation. In between, the loop gain reaches a point where the high and low frequency limitations cross, satisfying the gain magnitude condition for oscillation. Still, the feedback phase shift at this crossover must simultaneously reach 180° to produce oscillation.

To illustrate this gain and phase combination, Fig. 1.10 demonstrates the basic requirements for op amp oscillation. This figure grounds the normal signal input of the circuit to remove the effect of any applied signal upon the output voltage. With the input grounded, only the gain error signal $-e_o/A$ excites the input circuit. This signal must independently produce the output signal in order to sustain an oscillation. The circuit amplifies the $-e_o/A$ signal by the closed-loop gain A_{CL}, producing $e_o = A_{CL}(-e_o/A)$. In turn, this output signal reflects back through the op amp, producing the attenuated input signal $-e_o/A$. If this circuit gain and attenuation cycle, expressed in $e_o = A_{CL}(-e_o/A)$, balances, the circuit produces a self-sustaining oscilla-

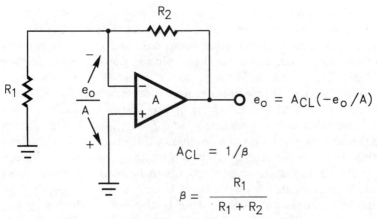

Figure 1.10 To sustain oscillation, error signal e_o/A and gain A_{CL} must support output voltage in the absence of an applied input signal.

tion. At higher frequencies, the roll off of A_{CL} prevents this oscillation condition.

At lower frequencies, $A_{CL} = 1/ß$, making the oscillation condition $e_o = A_{CL}(-e_o/A) = -e_o/Aß$. Then to sustain oscillation, the circuit must satisfy this new equality and only two solutions do, $e_o = 0$ and $Aß = -1$. The $e_o = 0$ solution indicates an oscillation of zero amplitude, representing the stable state. The $Aß = -1$ solution represents the actual oscillation state, as noted previously in the mathematical analysis of Sec. 1.3.1. However, this second solution places very specific magnitude and phase requirements upon the loop gain $Aß$. The condition $Aß = -1$ requires that $|Aß| = 1$ for the loop gain magnitude along with 180° of phase shift for the minus sign.

Consider the oscillation condition from another perspective beginning with the magnitude requirement. If $|Aß|>1$ at a given frequency, this indicates the presence of a larger open-loop gain A. Then the attenuated input error signal e_o/A, when amplified by a gain of $1/ß$, remains too small to support the required e_o. Only when the attenuating gain A equals the amplifying gain $1/ß$, does the circuit meet the magnitude condition for oscillation. Expressing this in an equation, $|A| = |1/ß|$ repeats the previous, mathematically derived condition for oscillation. Only at the intercept of the A and $1/ß$ curves do their magnitudes become equal to meet this condition. Then the circuit fulfills the magnitude condition for oscillation.

Next consider the phase requirement for oscillation. If $Aß$ lacks 180° of phase shift, then the minus sign of the $Aß = -1$ condition remains unsatisfied, preventing oscillation. Further, oscillation only results when this phase condition coincides with the magnitude condition mentioned. An $Aß$ phase shift of 180° at frequencies other than the intercept frequency does not produce oscillation. As described, the circuit fails to meet the magnitude condition for oscillation at those other frequencies.

Composite amplifiers permit a graphical illustration of this combined oscillation requirement. These amplifiers inherently produce the 180° phase shift required for the minus sign. They consist of two op amps connected in series within a common feedback loop, and each amplifier contributes a -20-dB per decade slope to the composite open-loop gain. As shown in Fig. 1.11, the two poles produce a -40-dB per decade gain slope, indicating 180° of phase shift over most of the amplifier's useful frequency range. Thus composite amplifiers meet the phase condition for oscillation over a broad frequency range.

Over much of the same range, the composite amplifier provides high open-loop gain. This gain and phase combination might first suggest widespread stability problems. However, at lower frequencies,

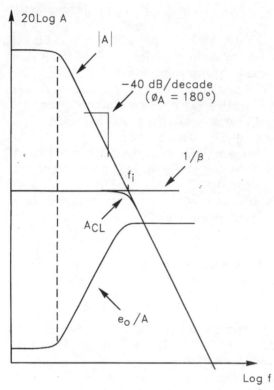

Figure 1.11 Composite amplifier response illustrating the fact that 180° of phase shift fails to support oscillation where e_o/A lacks sufficient magnitude.

the high open-loop gain actually serves to stabilize the circuit through the circuit's loop gain. High values for A increase the loop gain $A\beta$ to prevent the magnitude equality $|e_o| = |e_o/A\beta|$ required for oscillation. It does so by limiting the e_o/A error signal, as illustrated in the figure. At lower frequencies, high levels of open-loop gain A reduce this input error signal to a level insufficient to support oscillation. The e_o/A curve rises as gain A declines, but flattens when $A_{\rm CL}$ declines. The curve's rise must reach a certain level to support the oscillation condition of $e_o = -A_{\rm CL}(e_o/A)$. Only one point in the plot satisfies this oscillation condition. As described before, where $A = 1/\beta$, e_o/A reaches the level required to support oscillation. This intercept also marks the peak value for $A_{\rm CL}(e_o/A)$. Beyond there, $A_{\rm CL}$ rolls off with gain A, reducing e_o and leveling the e_o/A curve. With a level e_o/A curve, the $A_{\rm CL}$ roll off also rolls off the quantity $A_{\rm CL}(e_o/A)$ of the oscillation condition.

Before this intercept, e_o/A remains too small to support oscillation. After the intercept, the amplifier lacks the A_{CL} needed to sustain oscillation. Thus before or after the intercept, the 180° of feedback phase shift of the composite amplifier does not compromise stability. This phase shift produces oscillation only at the frequency of the intercept f_i. There gain magnitude conditions always permit oscillation given the inherent 180° phase condition of composite amplifiers. For practical applications of these amplifiers, phase compensation reduces the phase shift at f_i to restore frequency stability.[5,6]

The 1/ß intercept represents a critical mass point for frequency stability. There the magnitudes of the gain error and the feedback phase shift must both reach specific levels to support oscillation. Despite the very specific requirements for oscillation, the greatly varied applications of op amps make this critical mass condition all too easy to find. To contend with this, the 1/ß curve presents a visual prediction of the problem and provides insight into its solution. Phase compensation provides the solution as described in Chap. 3.

1.4 Op Amp Influence on Feedback Networks

The preceding chapter sections treat the feedback elements as independent impedances. This independence holds for moderate feedback impedances and typical op amp bandwidths. In other cases the op amp input impedance influences the feedback network and the related performance results described earlier. For accurate performance prediction, this influence must be included in the determination of the feedback factor. With higher feedback impedances, amplifier input capacitance significantly shunts the feedback, jeopardizing stability. In wider-bandwidth applications, the amplifier input inductance combines with the input capacitance, producing resonant peaking but not necessarily oscillation.

1.4.1 Effect of input capacitance

Op amp input capacitance significantly alters the feedback factor in circuits with higher feedback impedances. There the shunting of the capacitance alters the voltage divider action of the feedback network within the frequency response range of the circuit. This shunting affects both negative and positive feedback, but through different components of amplifier input capacitance, and different stability effects result from the shunting of the two feedback types.

First consider the effect of amplifier input capacitance upon nega-

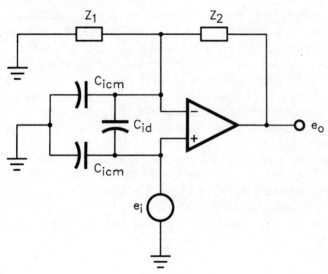

Figure 1.12 Op amp input capacitance alters feedback factor by shunting Z_1 of a negative feedback network.

tive feedback networks. Figure 1.12 illustrates this case with the three components of input capacitance drawn outside the amplifier. Amplifier input resistance components exist in parallel with these capacitances but these high resistances rarely alter the feedback factor. In practice, the capacitances dominate an op amp's input impedance starting at very low frequencies. Capacitive component C_{id} represents the differential input capacitance and appears between the op amp inputs. Two C_{icm} components represent the common-mode input capacitances present from either amplifier input to common. Two of the three capacitances shunt Z_1, altering the circuit's feedback factor. The upper C_{icm} connects directly across Z_1, and the C_{id} component does so through the e_i signal source. For the latter, the path through the low impedance of e_i completes the C_{id} return to ground, placing this capacitance in parallel with Z_1. The lower C_{icm} component of the figure remains isolated from the feedback network and does not affect the feedback factor. Thus the net capacitance shunting Z_1 equals $C_{id} + C_{icm}$, and that shunting reduces the fraction of the output e_o fed back to the op amp input, causing a decline in ß with increasing frequency. This decline represents a pole in the feedback loop and degrades the amplifier stability. Stability analysis reveals a rising 1/ß curve like the 1/ß$_2$ curve shown in Fig. 1.7, and phase compensation corrects this condition, as described in Chap. 3.

Amplifier input capacitance also shunts positive feedback net-

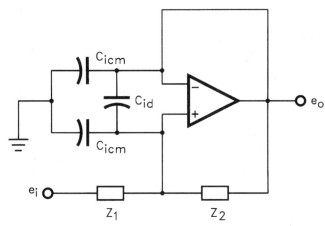

Figure 1.13 Input capacitance shunts both Z_1 and Z_2 of this positive feedback network.

works, but in ways that both aid and degrade stability. Figure 1.13 illustrates this shunting with positive feedback added to a voltage follower. There Z_1 and Z_2 supply feedback from the amplifier output to the amplifier's noninverting or positive input. This feedback raises the circuit's 1/ß curve, avoiding the stability compromise sometimes introduced by secondary amplifier poles. Chapter 3 describes this circuit in greater detail. In this case, the amplifier input capacitance alters the circuit's feedback factor in two ways through different capacitive components. Considering the three capacitances separately, the upper C_{icm} component of the figure connects between the amplifier output and ground. As such, this capacitance only presents a small capacitive load to the circuit and produces a negligible effect upon performance. However, the lower C_{icm} component of the figure now bypasses Z_1. Superposition analysis of ß demonstrates the effect of this bypass by grounding the e_i terminal, which places Z_1 in parallel with the lower C_{icm}. This bypass adds a pole in the positive feedback path, reducing ß$_+$ at higher frequencies.

However, this ß$_+$ reduction actually increases the circuit's net feedback factor as seen from feedback analysis. A combination of positive and negative feedback connections here complicates this analysis, but reflection upon the differential nature of an op amp's inputs restores simplicity. Those differential inputs subtract one input signal from the other, and thus, feedback signals supplied to the two inputs produce a differential effect. Similarly, the resulting negative and positive feedback factors subtract, producing a net feedback factor of ß = ß$_-$ −ß$_+$. Then the pole in the ß$_+$ response causes the circuit's net ß to increase rather than decrease. This makes the circuit's 1/ß curve decline rather than rise with increasing frequency, which in general signals good stability.

The final input capacitance component C_{id} counteracts this effect. Evaluation of Fig. 1.13 shows this component to be directly in parallel with Z_2, bypassing that impedance. This bypass produces the opposite effect from that described for the C_{icm} bypass of Z_1. The C_{id} bypass introduces a zero in the positive feedback and tends to produce a rising 1/ß curve. Thus two input capacitance components produce both a pole and a zero in the positive feedback path. If the zero prevails, degraded stability results. In such cases, adding capacitance across Z_1 makes the pole dominant to phase compensate the circuit. Stability analysis like that of Sec. 1.3 guides the determination of any phase compensation required.

1.4.2 Effect of input inductance

Typically, limited op amp bandwidth makes applications circuits insensitive to amplifier input inductance. Within the amplifier's open-loop response range, the impedance of this inductance typically remains too low to affect the amplifier performance. However, with very-high-frequency op amps, even the inductances of a package lead and an input wire bond become significant. Figure 1.14 shows the

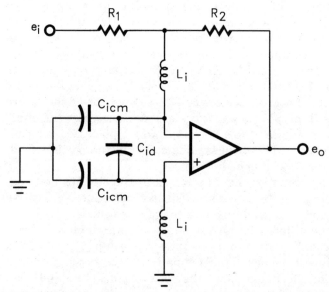

Figure 1.14 Op amp input inductance combines with input capacitance to produce a resonant tank circuit.

associated input inductances along with the amplifier input capacitances. There the various components of inductance and capacitance form a complex tank circuit. This tank circuit shunts R_1, producing an equally complex effect upon the feedback factor. To simplify the example, resistive feedback elements replace the general impedances used before. At its resonant frequency, the tank circuit shunts R_1 with zero impedance, reducing ß to zero as well. Fortunately, the amplifier's stability remains relatively immune to this resonance effect because it removes any feedback signal that could sustain oscillation. However, this resonance does increase distortion.

At first, graphical analysis of this resonance might seem to indicate oscillation. However, the 1/ß curve demonstrates stable conditions where typical gain and phase plots would suggest otherwise. Gain and phase measurements of the circuit in Fig. 1.14 produce the responses of Fig. 1.15, and both measurements detect the tank circuit resonance, as shown by the curves. Following unity crossover, the gain curve rises again above the unity axis, which typically suggests oscillation for lower closed-loop gains. Adding to the stability concern, the phase plot swings through 180° at the frequency of the gain peak. Also, the circuit's resistive feedback network suggests a flat 1/ß curve that would intersect this peak, indicating a lack of gain margin. Gain margin is the difference between the 1/ß level and the open-loop gain, where the loop phase shift reaches 180°. At this frequency, the level of the 1/ß curve must remain above that of the open-loop gain to avoid oscillation.

However, including the tank circuit in the feedback network shows that the actual 1/ß plot does not intersect the gain peak but merely rides over it, as shown. Without that intercept, oscillation cannot develop, regardless of the phase shift. The circuit lacks the loop gain required to support oscillation because the 1/ß curve shows that the feedback demand for loop gain rises in synchronization with the gain peak. In this example the gain margin remains high, as shown by the separation between the 1/ß and gain curves. This separation remains large throughout the region of higher phase shift and indicates good stability conditions. In a more common case, the gain peaking results from conditions in the amplifier output, rather than input. For that case no corresponding modification of feedback results and an intercept and oscillation result.

Even though the input inductance does not degrade stability in this example, this inductance does introduce a subtle distortion effect. The resulting resonant gain peak spawns others. Smaller gain peaks develop at the subharmonics of the resonant frequency due to the resonance interaction with distortion harmonics. If the resonance occurs at a frequency f_r, it reacts with the distortion harmonics of any signal

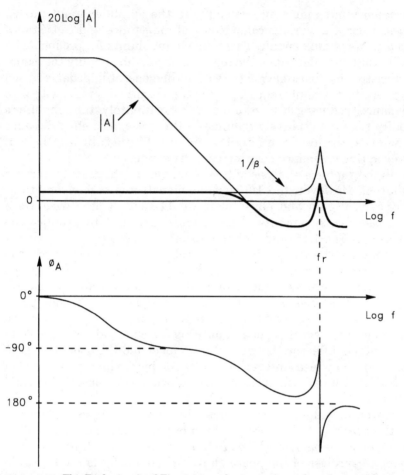

Figure 1.15 The Tank circuit of Fig. 1.14 produces a gain peak and a coincident, stability-preserving peak in the 1/ß curve.

at a frequency f_r/n, where n is any integer value. Any amplifier signal at f_r/n will have some distortion harmonic at f_r, and the resonant gain peaking amplifies this harmonic. The amplification of this component of the signal increases the overall amplitude of the output signal even though that signal has a frequency of f_r/n rather than f_r. If the increase is significant, open-loop gain measurements detect peaks at f_r/n. Fortunately, op amp loop gain typically minimizes the significance of these peaks in closed-loop applications.

References
1. J. Graeme, "Feedback Models Reduce Op Amp Circuits to Voltage Dividers," *EDN*, June 20, 1991, p. 139.

2. H. Black, "Stabilized Feedback Amplifiers," *Bell Syst. Tech. J.*, vol. 13, no. 1/35.
3. J. Graeme, "Feedback Plots Offer Insight into Operational Amplifiers," *EDN*, January 19, 1989, p. 131.
4. G. Tobey, J. Graeme, and L. Huelsman, *Operational Amplifiers: Design and Applications*, McGraw-Hill, New York, 1971.
5. J. Graeme, "Composite Amplifier Hikes Precision and Speed," *Electron. Des.*, June 24, 1993, p. 30.
6. J. Graeme, "Phase Compensation Perks up Composite Amplifiers," *Electron. Des.*, August 19, 1993, p. 64.

Chapter

2

Feedback Modeling and Analysis

Feedback modeling dramatically simplifies op amp circuit analysis. Given the model for a circuit, the circuit's performance analysis reduces to the determination of voltage divider ratios, and these ratios adapt standardized performance results to the specific circuit considered. Chapter 1 demonstrated the power of feedback analysis using the noninverting op amp configuration as an example. There the familiar Black feedback model permitted simple determination of the circuit's feedback factor ß to define the output errors produced by almost all op amp performance limitations. Following that, graphical analysis with the circuit's 1/ß curve defined bandwidth and stability conditions.

Unfortunately, Black's feedback model only applies to the noninverting circuit configuration of an op amp, leaving many other configurations unmodeled. Without knowledge of their feedback factors, these other configurations would require laborious circuit analysis on an individual basis. A more complete study of feedback theory would permit feedback modeling of these configurations, simplifying the analysis task. However, that formidable study exceeds the task required just for op amps. With but two inputs, the op amp presents a limited set of feedback alternatives. Instead, this chapter develops those feedback modeling concepts required for op amp circuits through a series of examples.[1,2] Specific connection examples demonstrate all of the possible variations to the feedback model for the op amp case. Demonstrating these, the examples that follow adapt the Black model to inverting circuits, positive feedback connections, and dual input connections.

Through these adaptations, a generalized modeling process evolves that will permit rapid analysis of any op amp configuration. Also, standardized performance results develop, which immediately apply to any modeled circuit. Then the results developed here and in Chap. 1 extend to all other op amp configurations. Each configuration modeled here exhibits a characteristic closed-loop response that parallels the noninverting example of Chap. 1. Through this parallel, the previous analysis results transfer directly to the modified models. All of the error, bandwidth, and stability results described before apply to circuit configurations represented by the new models developed here.

A generalized feedback model summarizes the modeling process and the standardized results. This generalized model directly represents most op amp circuits, and further extensions to it adapt the model to the most complex op amp circuits. Such extensions adapt to complex feedback connections that include multiple gain elements, gain in the feedback path, variable feedback, and multiple feedback connections. None of these conditions fit the generalized model developed here directly, but the determination of net circuit characteristics reduces these circuit complexities to the level of that model. This extends standard performance results to the most complex op amp circuits.

2.1 Inverting Configurations

In this section, the analysis of the inverting op amp configuration begins the adaptation of the Black model to other op amp configurations. This one-time analysis of the inverting circuit develops the circuit's transfer function for comparison with that of the model derived. Intuitive observations guide the model derivation, and then comparison of the model's transfer response with that of the circuit confirms the model's validity.

2.1.1 Feedback model for inverting configurations

The model modification for the inverting configuration begins with circuit analysis using Fig. 2.1. This figure shows the noninverting circuit discussed in Chap. 1 (Fig. 2.1a) and the inverting circuit considered here (Fig. 2.1b). Comparison of the two circuits initially reveals great similarity. The ground and e_i connections are simply interchanged between the two. However, this simple modification confuses the determination of the feedback factor. In the inverting case of Fig. 2.1b, the fraction of the amplifier output fed back to the input first appears to be near zero. The virtual ground, fundamental to the inverting configuration, assures near zero voltage at the op amp inverting input.

Feedback Modeling and Analysis

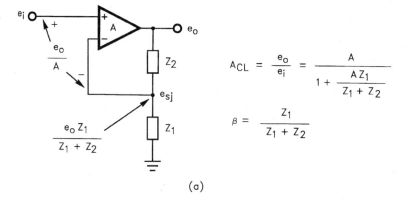

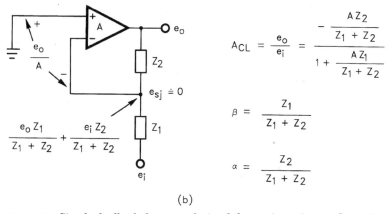

Figure 2.1 Simple feedback factor analysis of the noninverting configuration (a) repeats for the inverting configuration (b) when superposition analysis of e_{sj} separates the e_o effect from that of e_i.

However, this zero voltage condition simply results from counteracting signals delivered by e_o and e_i. The two signals both drive the inverting input, but from opposite ends of a feedback voltage divider. Superposition analysis of these signal drives shows that the signal at the amplifier inverting input or summing junction is

$$e_{sj} = \frac{e_o Z_1}{Z_1 + Z_2} + \frac{e_i Z_2}{Z_1 + Z_2}$$

The inverting configuration considered here ideally produces the closed-loop response $e_o = -(Z_2/Z_1)e_i$, making $e_{sj} = 0$. This leaves zero signal at the amplifier input, confusing the determination of the feedback factor.

However, the first term of the e_{sj} equation shows that $Z_1/(Z_1 + Z_2)$ remains the fraction of the output e_o fed back to the input. The second term of the equation merely expresses the independent feedforward effect of e_i. To avoid confusion, use superposition analysis to separate the two effects. Then, to determine the feedback factor, superposition sets all signals other than e_o to zero. This practice demonstrates that the feedback factor of an op amp circuit is the voltage divider ratio of the feedback network, regardless of the signals present in the actual circuit.

The switch from noninverting to inverting configurations also reflects in adjustments to the feedback model. Input signal connections for the inverting case differ in two ways, as illustrated in Fig. 2.2. This figure repeats the original Black model in Fig. 2.2a and shows the inverting configuration modifications in Fig. 2.2b. In Fig. 2.2b the input signal connects to a negative summation input and does so through an α transmission block. The classic feedback model of Fig. 2.2a connects e_i to a positive summation input with no transmission block. The sign change between the two corresponds to the associated signal connections at the op amp inputs. Here the input

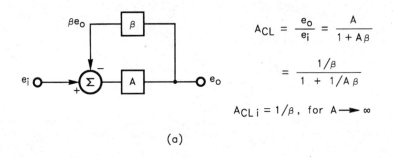

(a)

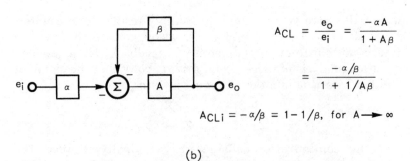

(b)

Figure 2.2 The feedback model for the noninverting configuration (a) converts to that of the inverting configuration (b) through a summation polarity change and addition of a feedforward α block.

signal drives the amplifier's inverting, rather than noninverting, input. The modified model accommodates this difference in Fig. 2.2b with a sign change at the summation input. In op amp feedback modeling, all summation inputs assume the same polarities as the corresponding amplifier inputs.

The added α transmission block models an added attenuation of the feedback network. The feedback model for the previous noninverting circuit, Fig. 2.2a, connects e_i directly to the summation element in correspondence with the direct signal connection of the circuit, Fig. 2.1a. However, the inverting circuit, Fig. 2.1b, connects e_i through the feedback network rather than through a direct connection. The new connection makes the feedback network attenuate the signal supplied to the op amp input, as reflected in the equation for e_{sj}. This equation expresses the signal reaching the amplifier input from e_i as an attenuated $e_i Z_2/(Z_1 + Z_2)$.

To account for this attenuation, the inverting feedback model of Fig. 2.2b inserts a feedforward factor α in the input path. This feedforward factor is the fraction of the applied input signal fed forward to the op amp input. As with the feedback factor, the feedforward factor results from a voltage divider ratio of the feedback network. For α, the divider ratio simply results from signal drive at the opposite end of the feedback network. Then, for inverting configurations,

$$\alpha = \frac{Z_2}{Z_1 + Z_2}$$

In practice, every input signal connection to a feedback model has a corresponding α, and direct input connections simply set α equal to unity.

2.1.2 Feedback analysis of inverting configurations

With the model adjustments described, the simplicity of feedback analysis extends to the performance evaluation of the inverting circuit. The feedback model of Fig. 2.2b sums the input and feedback signals for amplification by gain block A. This produces a model output of $e_o = A(-\alpha e_i - \beta e_o)$. Solving this equation for e_o/e_i yields the characteristic response of the model expressed in a closed-loop gain A_{CL}. Then manipulation of this result reduces it to a standard form with the denominator $1 + 1/A\beta$ and for the inverting configuration

$$A_{CL} = \frac{-\alpha/\beta}{1 + 1/A\beta}$$

This transfer response repeats the denominator of the noninverting analysis of Chap. 1 and transfers previous results to the inverting configuration. The noninverting bandwidth and stability results described before depend solely upon this denominator. Further, any different configuration that produces the same $1 + 1/A\beta$ response denominator shares those previous results.

This equation manipulation also transforms the numerator into a characteristic indicator. As mentioned in Sec. 1.2.2, the response numerator expresses the ideal closed-loop gain A_{CLi} of the configuration. That chapter section demonstrates the numerator and A_{CLi} equivalence for the noninverting case. In this inverting case, a comparison of response equations for circuit and model confirms this equivalence. Manipulation of the associated A_{CL} expressions transforms them for term-by-term comparison between the circuit and the model. In every case, the response denominator guides the manipulation of a given A_{CL} response equation, and all op amp A_{CL} equations can be reduced to a form with a denominator of $1 + 1/A\beta$. Then a comparison of terms translates circuit parameters into those of the model.

A comparison of responses for the circuit in Fig. 2.1b and the model in Fig. 2.2b first confirms the values defined previously for α and β. These parameters alone determine the ideal closed-loop gain. For the inverting configuration, the actual closed-loop gain for both the circuit and its model is

$$A_{CL} = \frac{-\alpha/\beta}{1 + 1/A\beta}$$

When the loop gain $A\beta$ is large, the denominator becomes unity and A_{CL} equals the numerator, the ideal A_{CLi}. Then large $A\beta$ values reduce A_{CL} to the familiar inverting circuit result of $A_{CLi} = -\alpha/\beta = -Z_2/Z_1$.

The magnitude of this closed-loop gain falls below the $(Z_1 + Z_2)/Z_1$ of the noninverting case without a corresponding increase in bandwidth. The feedback factor, rather than the closed-loop gain, sets the circuit bandwidth. The inverting configuration retains the feedback factor of the noninverting configuration even though the gain magnitudes differ for the two cases. As described, both configurations produce a feedback factor of $\beta = Z_1/(Z_1 + Z_2)$. In the previous noninverting case, the $1 + 1/A\beta$ denominator of A_{CL} places the 3-dB bandwidth at βf_c. This denominator repeats in the A_{CL} expression for the inverting case, and the frequency response discussions of Sec. 1.2 again apply. Thus the bandwidth remains at βf_c and the gain-bandwidth product drops in moving from the noninverting to the inverting configuration.

The frequency stability results of the noninverting case also transfer to the inverting case through the common response denominator. Previously, Sec. 1.3 developed these results based upon the $1 + 1/A\beta$ denominator repeated here for the inverting case. When $A\beta = -1$, this denominator again makes A_{CL} infinite to support oscillation. This $A\beta$ condition occurs at the $1/\beta$ intercept with open-loop gain A if the loop phase shift equals 180° at that point. There the rate of closure of the gain and $1/\beta$ curves predicts the loop phase shift, as also described in Chap. 1.

This BW = βf_c relationship and the stability conditions extend to all other op amp configurations as well. In fact, the transfer response of almost any feedback system can be written in a form that produces the $1 + 1/A\beta$ denominator. Then the bandwidth and stability conclusions previously drawn from this denominator apply to all responses expressed with this denominator. Also, when written in this standard form, the response equation always reflects the ideal closed-loop gain as the response numerator. As described in Sec. 1.2.3, the generalized transfer response for any op amp configuration is simply

$$A_{CL} = \frac{A_{CLi}}{1 + 1/A\beta}$$

The feedback modeling that follows reduces each configuration's transfer response to this generalized form. Then previous conclusions drawn from the standard equation translate directly to these configurations. Each configuration links directly to the bandwidth and stability conclusions of the initial, noninverting example of Chap. 1. The only variable factor in this process is A_{CLi}, which expresses the fundamental response difference between op amp circuit configurations. For each A_{CLi} some combination of α and β defines an ideal closed-loop gain unique to the associated configuration. In each case, writing the response of the feedback model in the standard form defines A_{CLi}.

Similarly, the error gain of op amp circuits A_{CLe} retains the generalized form described in Sec. 1.2.4. For this gain no equation difference separates the different amplifier configurations. This gain amplifies all of the input-referred error signals of e_{id} in Fig. 1.2. As described there, the basic error gain is $1/\beta$. This gain and feedback factor relationship extends to A_{CLe} for all op amp configurations independent of a configuration's closed-loop gain. Amplifier frequency response limitations modify this basic gain at higher frequencies. Here again the result remains independent of the op amp configuration. After the $1/\beta$ curve intercepts the amplifier open-loop gain curve, the op amp error gain rolls off, following the op amp's open-loop response. This defines the error gain as

$$A_{CLe} = \frac{1/\beta}{1 + 1/A\beta}$$

From the discussion in Sec. 1.2.3 the A_{CLe} roll off starts at the closed-loop bandwidth limit of the circuit at BW = βf_c.

2.1.3 Noise bandwidth and the inverting configuration

To reduce noise, inverting op amp configurations often include a feedback bypass capacitor that reduces high-frequency gain. However, this solution treats signal and amplifier noise sources differently, often resulting in higher than expected output noise. Prediction of the resulting noise performance depends upon the distinction between closed-loop gain A_{CL} and error-signal gain A_{CLe}.

Figure 2.3 illustrates the C_f bypass added for noise bandwidth reduction. Two noise voltage sources affect the noise performance of the circuit through the noise content of input signal e_i and the amplifier's input voltage noise source e_n. To reduce noise, capacitor C_f bypasses resistor R_2, rolling off the closed-loop gain with a pole at $1/2\pi R_2 C_f$. This successfully reduces circuit bandwidth for the noise contained in signal e_i but only partially restricts the output noise due to e_n. The error gain A_{CLe}, not the closed-loop gain, amplifies e_n. Figure 2.4 illustrates the A_{CL} roll off along with the A_{CLe} response. As expected, A_{CL} rolls off with a continuous, single-pole response. However, A_{CLe} displays a more complex response that reduces the amplification of the amplifier's input noise signal e_n only partially. As shown, this noise gain equals 1/ß prior to the 1/ß intercept with the gain magnitude curve. After that, A_{CLe} rolls off with the amplifier

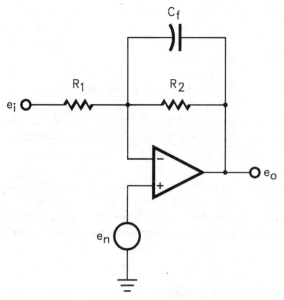

Figure 2.3 Capacitive bypass of R_2 short circuits the inverter noise gain presented to e_i but only reduces this gain to unity for amplifier noise e_n.

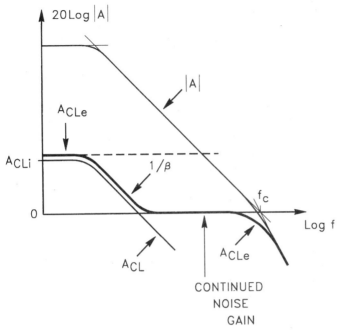

Figure 2.4 Graphical analysis of two noise bandwidths of Fig. 2.3 reveals that 1/ß continues high-frequency gain for amplifier noise e_n.

open-loop response. The addition of C_f adds a pole to the 1/ß gain, and this pole also occurs at $1/2\pi R_2 C_f$.

Initially, this pole rolls off 1/ß, as shown in the figure, but a zero interrupts the roll off. The zero produces continued noise gain out to the open-loop roll off of the amplifier. Analysis of 1/ß shows the zero to be at $1/2\pi(R_1 || R_2)C_f$. This zero results from the fact that the feedback factor cannot be raised above unity. At that unity level, the fraction of the output fed back to the input reaches 100%. When C_f completely bypasses R_2, the capacitor produces short-circuit feedback for the amplifier of Fig. 2.3, and this connects the amplifier as a voltage follower, as seen by the noise signal e_n in Fig. 2.3. Thus the A_{CLe} noise gain drops to unity in Fig. 2.4 and continues at this level out to the open-loop roll off of the op amp. Dropping the 1/ß curve to the unity axis does reduce the noise gain received by e_n, but the result can be deceptive. At higher frequencies, the addition of C_f drops this gain from the dashed line of Fig. 2.4 to the unity axis. However, the response curves do not provide an accurate visual assessment of the degree of improvement. The unity-gain region of A_{CLe} often covers the majority of the actual amplifier bandwidth. Here the logarithmic scale of the frequency axis deceives visual perception of the relative significance of this region. Sometimes this unity-gain region contributes the majority of the circuit's output noise. Only quantitative analysis or measurement indicates the actual degree of improvement.

For further noise improvement, the op amp's open-loop bandwidth must be restricted. Then the open-loop roll off interrupts the continued noise gain at a lower frequency. Reduced amplifier bandwidth results from simply selecting a slower amplifier or from adding phase compensation. Most op amps lack the provision for added phase compensation, but Chap. 4 describes an external compensation alternative. Shown in Fig. 4.3, this phase compensation normally counteracts the effect of a capacitance load,[3] but it also restricts the noise bandwidth supplied to e_n.

The leveling off of 1/ß in this inverting case also shows why the addition of C_f imposes a unity-gain stability requirement upon the op amp. Based upon graphical analysis, it might first be thought that the roll off produced by C_f avoids stability problems by reducing the closed-loop gain at high frequencies. However, the 1/ß curve, not the closed-loop gain curve, predicts frequency stability. The addition of C_f makes the 1/ß curve produce a critical intercept with the open-loop gain curve at f_c, and this intercept requires a unity-gain stable amplifier. As a result, better frequency stability often results without C_f. Then the 1/ß curve continues along the dashed line shown for an intercept with gain A well removed from the added phase shift introduced by f_c.

2.2 Positive Feedback Configurations

Positive feedback normally suggests circuit latching or oscillation. With op amps, however, this feedback adds options to linear, stable circuits. In fact, positive feedback can improve rather than degrade circuit stability. The key lies in a combination of positive and negative feedback, where a dominant negative feedback ensures the stable operation normally associated with that feedback. The combined feedback alters the circuit bandwidth, the stability conditions, and the feedback model. An example circuit illustrates these effects for the general case of positive feedback.

2.2.1 Positive feedback example

Figure 2.5 shows an op amp configuration with feedback to both of the op amp's inputs. There the op amp output connects directly to the inverting input, producing the unity negative feedback of a voltage follower. The output also connects to the noninverting amplifier input through a feedback network, and the voltage divider attenuation of

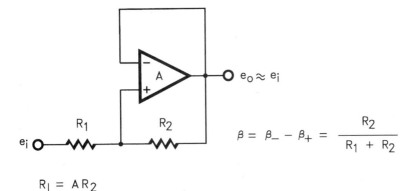

$R_l = AR_2$

Figure 2.5 Addition of positive feedback reduces the feedback factor of a voltage follower and reveals that positive and negative feedback factors subtract in the net feedback factor.

this network makes the associated positive feedback less than unity. Thus the unity negative feedback has the greater magnitude and prevails in the feedback control of the circuit.

To illustrate the combined feedback option, consider the example circuit shown. There the combination of positive and negative feedback permits voltage-follower operation with an op amp having reduced phase compensation. Such op amps routinely offer greater bandwidth and slew rate to higher-gain op amp applications. Positive feedback extends the greater slew-rate advantage to lower-gain applications. Operating at the lower-gain extreme, a conventional voltage follower has unity feedback factor and must normally have an op amp that is compensated for unity-gain stability. Op amps with reduced phase compensation have a minimum stable gain greater than unity and normally cannot be used in this application.

However, modification of the circuit's feedback factor makes the higher slew rate of the reduced phase compensation option available to the voltage follower. This typically improves the slew rate by a factor equal to the minimum stable gain A_{min} specified for the op amp. In the example, positive feedback reduces the feedback factor without altering the closed-loop gain. Then the reduced feedback factor simply raises the 1/ß curve in the stability analysis. As described in Sec. 1.3, the frequency stability of an op amp configuration depends upon the feedback phase shift developed at the frequency of the intercept of the A and 1/ß curves. Figure 2.6 shows these curves for the reduced phase compensation and added positive feedback of the example considered here. Because of the reduced compensation, the open-loop gain curve A exhibits two response poles above the unity-gain axis, producing a −40-dB per decade response slope before the curve reaches the unity-gain or

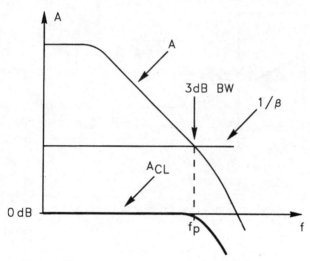

Figure 2.6 The positive feedback of Fig. 2.5 raises the 1/ß curve to a region of stable intercept above the second pole of a two-pole, open-loop gain responses.

0-dB axis. This slope corresponds to 180° of phase shift and indicates oscillation for a 1/ß intercept at unity gain. Normally, that specific intercept would occur with a voltage follower because then 1/ß = 1.

However, the positive feedback of the example reduces the net feedback factor to raise the 1/ß curve as shown. This moves the intercept to a region of reduced open-loop gain slope and assures frequency stability. Raising the 1/ß curve also moves the intercept back in frequency, reducing the closed-loop bandwidth. However, the resulting bandwidth still equals that of the unity-gain compensation alternative. In that alternative, adjusting phase compensation, instead of 1/ß, reduces the bandwidth by the same amount encountered here. For greatest bandwidth with Fig. 2.6, the intercept is set at the level of the amplifier's minimum stable gain. This results for 1/ß = A_{min}, where A_{min} is the minimum stable gain specified for the amplifier. Then the positive feedback circuit provides a factor of A_{min}:1 greater slew rate than a conventional voltage follower built with an equivalent unity-gain compensated op amp.

To permit this feedback setting, ß must be determined for the circuit. Once again, the feedback factor definition and the basic feedback model fail in this task. The dual feedback paths confuse the determination of the fraction of the output fed back to the input. Also, the classic feedback model, presented earlier, models only one feedback path. This model requires modification to accommodate the added positive feedback of this case.

2.2.2 Modeling positive feedback

Extension of feedback modeling to the positive feedback case follows the process described earlier for the inverting configuration. First, circuit analysis determines the closed-loop gain expression. Then modifications to the basic feedback model adapt it to the new circuit configuration and analysis determines the closed-loop gain of the resulting model. Comparison of the two closed-loop gain expressions interprets the model parameters in terms of circuit elements. Finally, manipulation of the closed-loop response equation converts it to standard form, transferring previous analysis results to the new configuration.

Figure 2.7 shows the generalized circuit representing the positive feedback example circuit, and analysis of this circuit indicates various α and ß factors. First, the direct output-to-input connection of the negative feedback sets $\beta_- = 1$. Then loop analysis defines the positive feedback signal and the closed-loop response as indicated in the figure. The positive feedback expression at the op amp noninverting input displays familiar voltage divider ratios linked to e_o and e_i. There $Z_1/(Z_1 + Z_2)$ reflects the positive feedback factor and $Z_2/(Z_1 + Z_2)$ reflects an input feedforward factor. These voltage divider ratios also appear in the A_{CL} expression of the figure, and they link the feedback and feedforward factors to the closed-loop gain.

To complete this link, feedback modeling adapts the basic feedback model to the features of the example circuit. This process illustrates the general modeling of positive feedback, and Fig. 2.8 shows the resulting feedback model. There two adjustments adapt the basic model of Fig. 2.2a to the positive feedback case. First, the revised

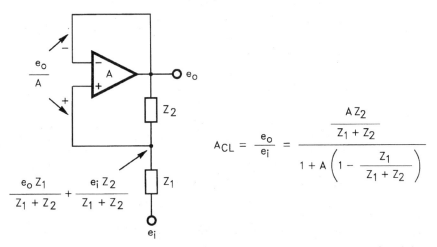

Figure 2.7 Redrawn circuit for Fig. 2.5 highlighting voltage divider actions that define α and ß factors for circuit.

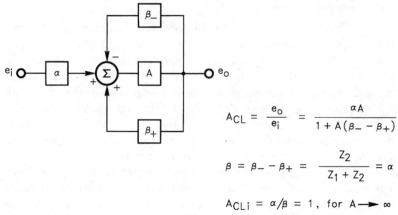

Figure 2.8 Feedback modeling of the positive feedback case defines net feedback factor as the difference between negative and positive feedback factors.

model adds a feedforward factor δ in series with the signal input. This follows from the earlier, inverting amplifier modeling and accounts for the connection of e_i to a feedback network, rather than directly to an op amp input. The signal coupled through this network reaches the amplifier input attenuated by the factor α. Note that e_i couples to plus inputs on both the op amp of the circuit and the summation element of the model.

The second model change inserts the β_+ feedback path shown, adapting the model for any positive feedback application. The β_+ feedback block connects to a plus input on the summation element, whereas feedback through the β_- block remains connected to a negative summation input. Then all summation polarities match those of the corresponding op amp input connections in the circuit. Now the question remains as to how the negative and positive feedbacks combine to control the circuit response.

2.2.3 Positive feedback and the feedback factor

The effects of positive and negative feedback combine for op amp circuits to develop a net feedback factor. This combination results from the differential operation of the op amp input circuit. That circuit makes the op amp respond to the difference between the signals at the amplifier's two inputs. Thus the op amp subtracts any two feedback signals connected to opposite amplifier inputs. This subtraction repeats in the model discussed through opposite summation polarities applied to the two feedback signals.

Subtraction by the differential inputs produces the net feedback factor for the combined feedback case, and that factor equals the difference between negative and positive feedback factors. Analysis of the model shows its response to be

$$A_{CL} = \frac{\alpha A}{1 + A(\beta_- - \beta_+)}$$

This response combines the feedback factor effects in a net feedback factor of ß = $\beta_- - \beta_+$. Note that the modeling does not make this net feedback factor $\beta_+ - \beta_-$ since that would insert a minus sign in the denominator of the equation. Then the A_{CL} equation would no longer be of the standard form, signaling a modeling error.

Comparison of this model response with that of the circuit confirms the earlier expressions for the α and ß terms. Rewriting the response expression translates the denominator to the standard form 1 + 1/Aß. Then

$$A_{CL} = \frac{\alpha/\beta}{1 + 1/A\beta} = \frac{A_{CLi}}{1 + 1/A\beta}$$

where ß = $\beta_- - \beta_+$. In this form the A_{CL} denominator transfers the Chap. 1 bandwidth and stability conclusions to the positive feedback case. Also, the A_{CL} numerator identifies the circuit's ideal closed-loop gain as $A_{CLi} = \alpha/\beta$.

While based on a specific example, this A_{CL} result holds for all op amp circuits having the configuration modeled. This avoids the need for detailed analysis of each specific circuit. Once the A_{CL} equation comparison confirms the model for the circuit configuration, the model's results remain the same for all circuits of that configuration. This reduces the analysis of specific circuits to the determination of α and ß factors. As explained before, the simple voltage divider ratios of feedback networks define these factors. The divider ratio is unity for the direct output-to-input connection of the example's β_- feedback. However, the model derived also holds for other cases where β_- is set by a feedback network.

For the specific circuit of the example the desired voltage-follower response results, but circuit errors are amplified. To determine the ideal gain, analysis expresses the α and ß factors in terms of actual circuit elements. Reading the individual factors from the circuit in Fig. 2.7 and subtracting the two ß factors gives

$$\beta = \frac{Z_2}{Z_1 + Z_2} = \alpha$$

Then $A_{CLi} = \alpha/\beta = 1$ for the desired voltage-follower response. However, with ß<1, the input errors of the amplifier are now amplified by 1/ß>1. Previously, stability considerations defined ß for the example circuit through the requirement that $1/\beta = A_{min}$. Also, previous discussion indicated a slew-rate improvement by a factor of A_{min} for this example. Thus the example circuit amplifies input errors by about the same factor by which it improves the slew rate. Typically, this is a factor of 5 to 10. The error signal gain rolls off with the amplifier open-loop response, as expressed in the equation for A_{CLe}. Also, adding a capacitor in series with R_2 in Fig. 2.7 blocks the increased error gain at low frequencies, avoiding the amplification of input offset voltage.

The positive feedback of the example also reduces the circuit's input impedance, but by far less than first expected. At first the input impedance of the circuit appears greatly reduced since the input signal drives a feedback network. Normally a voltage follower presents the very high impedance of an op amp input to the signal source. In contrast, connecting the signal source to the feedback network of an inverting circuit meets with an input impedance equal to the input resistor. This great difference in input impedance would also result here, except for the bootstrap action of the positive feedback. The follower action of the circuit keeps both ends of the feedback network at almost the same signal level. The only signal across Z_2 in Fig. 2.7 is the small e_o/A gain error signal. Thus the feedback network draws very little signal current from the signal source. The resulting input impedance becomes $R_I = AZ_2$, and the op amp's open-loop gain boosts the input impedance to a high level.

2.3 Dual Input Connections

The previous section illustrates the use of both amplifier inputs for feedback connections. Other op amp configurations use the two inputs for input signal connections. Input modifications to the feedback model accommodate these configurations, following the modeling process demonstrated before. The two input signal connections can be to the same signal source or to separate ones. In the simplest case, one signal source couples to both op amp inputs and this serves to illustrate the model modifications.

2.3.1 Dual input example

The example of Fig. 2.9 illustrates dual coupling of one signal source to the two op amp inputs. This circuit selectively amplifies the op amp input error signal for greatly improved resolution of error measurement. Distortion measurement in Chap. 6 particularly benefits from this selective gain. Examination of the circuit reveals the selective gain supplied to the gain error signal e_o/A. Feedback supports

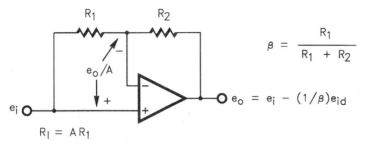

Figure 2.9 Bootstrapped feedback selectively amplifies error signal e_{id} for measurement while retaining unity gain for input signal e_i.

this signal across resistor R_1, developing a feedback current of e_o/AR_1. This current then flows through R_2, developing an amplified replica of e_o/A at the op amp output. This operation reflects an error signal gain of $(R_1 + R_2)/R_1$, and this gain equals $1/ß$, as seen by reading ß from the voltage divider ratio of the feedback network.

The circuit excludes test signal e_i from this amplification by isolating that signal from R_1. Through its bootstrap connection, R_1 rides on top of signal e_i, and this signal merely shifts the two op amp input voltages without directly developing a voltage on R_1. In the figure, the e_i signal drives the noninverting input, and feedback forces the inverting input to follow this signal also, making e_i a common-mode signal for R_1 and producing very little differential signal across the resistor. Thus R_1 conducts little e_i related feedback current to R_2, producing a similarly small e_i related voltage drop on R_2. As a result, the op amp output essentially follows e_i, and signal e_i receives only unity gain from the circuit, whereas the amplifier error signal receives a gain of $1/ß$. At the amplifier output, this selective gain raises the error out of the background level of the test signal. Thus the bootstrap connection reduces dynamic-range requirements for the error measurement by the amount of the selective gain.

From another perspective, the bootstrap increases the dynamic range of the input test signal. Returning R_1 to common, instead of bootstrapping, does not change the gain received by the e_o/A error signal. However, this alternate return produces the same gain for e_o/A and e_i. Then the gain applied to e_i reduces the maximum level this signal can reach without causing output saturation. With bootstrap, the unity gain presented to e_i allows this signal to span the full voltage range of the op amp input capability without causing output saturation. This permits exercising the full common-mode range of the op amp for testing common-mode-rejection related errors.

However, this selective gain also reduces the feedback factor, and thereby bandwidth, as described for the last example. In distortion measurement, the resulting bandwidth must be predicted accurately to determine the number of higher-frequency harmonics amplified. Examination of this bandwidth reduction, from a circuit perspective, adds intuitive feel to the previous graphical analysis. In the ideal, infinite open-loop gain case, the circuit in Fig. 2.9 produces $e_o = e_i$. Including the input gain error signal of the amplifier reduces the actual Fig. 2.9 output $e_o = e_i - e_o/Aß$. As gain A declines with frequency, the output signal reduction increases. At some point, the drop in output reaches the -3-dB point of the bandwidth limit. The reduced feedback factor ß increases the error signal $e_o/Aß$ directly, producing the -3-dB point at a lower frequency.

The performance described for the example circuit here closely resembles that of the preceding positive feedback example. Both circuits maintain unity gain to the signal source but operate the amplifier with ß < 1. Previously, the reduced feedback factor permitted less amplifier phase compensation but resulted in greater gain to the error signals. Here the circuit intentionally adds gain to the error signal for measurement applications. However, the only real difference between the two circuits lies in the applications described. In practice, the circuits realize the same results through different circuit configurations, and the two circuits are interchangeable from an applications standpoint.

2.3.2 Modeling dual input connections

The preceding two circuits both produce a voltage-follower response, but they differ in the feedback modeling that the circuits illustrate. The previous positive feedback circuit illustrates dual feedback, whereas the selective amplification circuit here demonstrates dual input connections. To model the dual input case, Fig. 2.10 first presents the example circuit in the generalized form used before. In the figure, the input signal e_i couples to the inverting amplifier input through the feedback network, and this connection produces an α feedforward factor equal to the voltage divider ratio $Z_2/(Z_1 + Z_2)$. The same network produces a ß feedback factor equal to the voltage divider ratio seen from the opposite end of the network. This simply switches the numerator impedance of the ratio and ß = $Z_1/(Z_1 + Z_2)$. Loop analysis defines A_{CL} for the circuit, and manipulation of the result expresses this gain in terms of the α and ß factors described. The A_{CL} expression of the figure reflects the result and compares directly with the model result that follows.

Figure 2.11 shows the feedback model for the dual input example. There signal e_i connects to the summation element through two

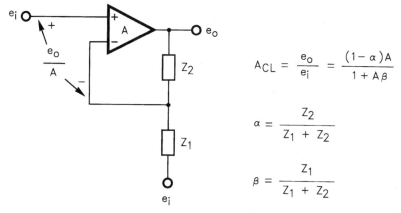

Figure 2.10 Redrawing Fig. 2.9 illustrates α and ß factors for the circuit through a simple voltage divider viewed from both ends.

paths. First the signal connects directly to a plus input, and this represents the circuit's direct connection to the op amp's noninverting input. This model connection presents an α of unity. However, in other application circuits, this noninverting input connection couples through a resistor network, making α less than unity. The second input connection of the model illustrates this case for an inverting input connection. There signal e_i connects through an α block representing the attenuation of the feedback network. This connection drives a minus summation input, corresponding to the inverting input connection of the circuit feedback. Otherwise the gain and feedback blocks of the model remain those of the basic feedback model.

Analysis of the completed model reproduces the A_{CL} response of the preceding figure and confirms the α and ß factors defined there from

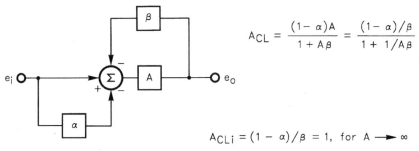

Figure 2.11 Feedback model for dual input example of Fig. 2.10 adds a second input connection through an α attenuation block.

the circuit diagram. Deriving and manipulating the model's response expression translates it to the standard form as

$$A_{CL} = \frac{(1-\alpha)/\beta}{1 + 1/A\beta} = \frac{A_{CLi}}{1 + 1/A\beta}$$

This result repeats the denominator $1 + 1/A\beta$ common to all of the previous results, and the bandwidth and stability conclusions previously drawn also apply to the dual input configuration of the example. Once again, $BW = \beta f_c$ defines closed-loop bandwidth, and the intercept of the A and $1/\beta$ curves communicates frequency stability conditions. Also the numerator of this standard form expression again defines the ideal closed-loop gain, making $A_{CLi} = (1-\alpha)/\beta$. Substitution of α and β expressions here shows that $A_{CLi} = 1$, as expected from the earlier discussion.

2.4 Feedback Analysis Process

The feedback modeling principles of the preceding examples readily extend to any op amp application. Then simple model analysis simultaneously defines many performance characteristics of the application circuit, avoiding the more complex response analysis of the actual circuit. When applying feedback modeling, the circuit itself is only analyzed when questions arise about the validity of the feedback model drawn.

A given feedback model covers all application circuits sharing the circuit's general configuration. Thus one model analysis produces results covering many specific circuits. Also the number of possible feedback models remains limited by the two inputs of the op amp. Feedback and input signals drive these two inputs in a limited number of possible configurations. Thus feedback analysis consolidates op amp circuits into a limited set of feedback models. For each model, the solution of one loop equation produces generalized performance results for the corresponding circuit configuration. Substitution of a given circuit's α and β factors in these results adapts the generalized results to the specific circuit. In this way, feedback modeling truly reduces op amp circuit analysis to the determination of voltage divider ratios. This section illustrates the power of feedback analysis beginning with a summary of the process. Then an application of the process to a complex circuit demonstrates the simplicity of this approach.

2.4.1 Summary of the feedback analysis process

The examples of preceding sections illustrate a variety of performance indicators and feedback modeling principles. Together, these indica-

Feedback Modeling and Analysis   **49**

tors and principles unify the analysis process for all op amp circuits. Three steps complete the process through determining the α and ß factors, drawing the model, and finding the model's transfer response. The previous examples illustrate these steps and the modeling principles summarized here. Redrawing the circuit eases the first two steps. Drawing the feedback networks as simple voltage dividers clarifies the attenuations applied to feedback and input signals. Then these dividers quickly display the circuit's α and ß factors as the corresponding voltage divider ratios. Output and input signals drive a given network from opposite ends, resulting in different divider ratios, and superposition analysis separates the effects of the two signals in the ratio determinations.

At this point in the analysis process the net feedback factor of the circuit already communicates a great deal of performance information. Most circuits use only negative feedback, and the net feedback factor equals the ß factor of this feedback. Other circuits use both negative and positive feedback, resulting in $ß_-$ and $ß_+$ factors, and produce a net feedback factor of $ß = ß_- - ß_+$. In either case, the circuit's net feedback factor defines performance errors, bandwidth, and stability conditions as described in Secs. 1.1 through 1.3.

In the next step of the process, construction of the feedback model also follows from the redrawn circuit. There voltage dividers connected to the op amp inputs indicate α and ß attenuator blocks linking those inputs to the overall circuit inputs and output. These links are repeated in drawing the feedback model. Direct connections sometimes replace the voltage dividers, and there the direct connections are repeated in the model. Each attenuator or direct connection supplies an input signal to the summation element of the model. There polarity assignments for each input signal correspond to the op amp input polarities in the circuit. Negative inputs on the model summation element correspond to circuit connections at the op amp's inverting input. Similarly, positive inputs to the element correspond to connections at the op amp's noninverting input.

Finally, analysis of the model defines the closed-loop response of the circuit. This requires only one loop equation describing e_o in terms of the model's summation result times the open-loop gain A. Solving this equation for $A_{CL} = e_o/e_i$ defines the transfer response of the circuit, and manipulation of this result arranges it in a standard form with a denominator of $1 + 1/Aß$. Then the numerator of this A_{CL} response equals the ideal closed-loop gain A_{CLi}. In addition to providing insight, this standard form requirement helps detect analysis and modeling errors. If the denominator cannot be reduced to $1 + 1/Aß$, an error exists.

2.4.2 Defining the net feedback factor

To illustrate the power of this analysis process, consider the voltage-controlled current source of Fig. 2.12. This op amp connection produces a voltage-controlled output current, and positive feedback removes the circuit's load voltage sensitivity. Voltage developed upon load R_L acts as a signal to the op amp's noninverting input, and circuit amplification of this signal adjusts the op amp output voltage to accommodate the load voltage effect. The added output voltage supplies a correction current through the positive feedback network formed by resistors R_2/n and R_2 in conjunction with the load R_L. The current supplied by this network to R_L accurately compensates for the effect of the load voltage once the two 1:1/n resistor ratios illustrated are established.

However, the dual feedback connections of this circuit obscure any intuitive anticipation of its performance characteristics, and the circuit structure offers little insight into circuit bandwidth and the effects of input error signals. Further, the positive feedback used by the circuit raises the question of frequency stability, and the dynamic range restriction imposed by the amplifier output voltage limit is not apparent. Straightforward analysis of all of these performance characteristics represents a formidable task.

Feedback modeling reduces the task to one loop equation through the information derived from the feedback factor and the closed-loop response. The resulting feedback model represents all op amp circuits having feedback to both op amp inputs. Redrawing the circuit in Fig.

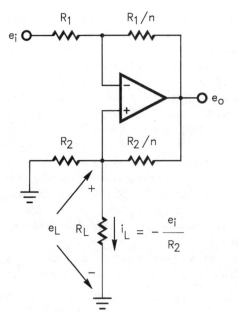

Figure 2.12 Multiple feedback connections of a common current source circuit inhibit intuitive anticipation of circuit performance.

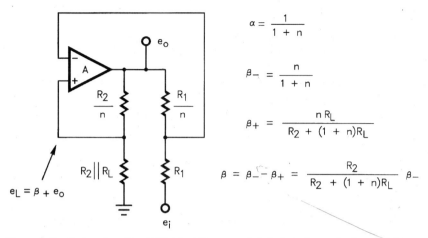

Figure 2.13 Redrawing Fig. 2.12 identifies α and ß factors through simple voltage dividers.

2.13 begins the analysis process. There simple voltage dividers represent the feedback networks for ease of analysis. These dividers supply the op amp inputs with attenuated signals from the amplifier output and from the signal input. As shown, combining R_2 and R_L reduces the positive feedback network to a simple divider with one divider element equaling $R_2 || R_L$.

Then simple divider voltage ratios define the α and ß factors for the circuit, and this yields immediate insight into the circuit's performance. For the circuit of Fig. 2.13,

$$\alpha = \frac{1}{1+n}$$

$$\text{ß}_- = \frac{n}{1+n}$$

$$\text{ß}_+ = \frac{nR_L}{R_2 + (1+n)R_L}$$

Then ß = ß_− − ß_+ defines the net circuit feedback factor as

$$\text{ß} = \frac{R_2}{R_2 + (1+n)R_L} \text{ß}_-$$

2.4.3 Intermediate results

This feedback factor result alone predicts the circuit errors, bandwidth, and frequency stability. As described in Sec. 1.1, the net feedback factor of an op amp circuit defines the circuit gain for the numerous input-referred errors of the op amp. There the input-referred error signal e_{id} represents the combined error effects of the major op amp performance limitations. Op amp circuits amplify e_{id} by a gain of 1/ß to produce an error e_{id}/ß at the op amp output. From this op amp output, the resulting error feeds back to the actual output terminal of this current source example. The amplified error feeds back to load R_L through the attenuation factor ß$_+$ of the positive feedback network. This attenuated signal develops an error current in R_L of (ß$_+$/ß)(e_{id}/R_L). Typically, ß$_+$/ß is large, and this amplifies the effect of e_{id} upon the load current.

The net feedback factor so determined also predicts the circuit's bandwidth through the results of Sec. 1.2. There analysis defines the bandwidth of typical op amp circuits as ßf_c, where f_c is the frequency of the amplifier unity-gain crossover. The positive feedback of the example considered here reduces the bandwidth by reducing the net circuit ß. From the preceding equation for ß, the positive feedback makes the net feedback factor a fraction of the negative feedback factor ß$_-$. This fraction depends upon the load resistance R_L and approaches zero as R_L becomes large, making the bandwidth also approach zero.

The ß equation also reveals stability information, both by itself and in conjunction with traditional graphical analysis. By itself, the equation resolves the question of oscillation resulting from the positive feedback. Examination of this equation shows that ß retains a positive polarity for all values of R_L, and this assures that negative feedback prevails regardless of the load resistance. Otherwise the positive feedback would dominate the circuit to cause oscillation or latching. Further stability information results from graphical analysis using the ß result in feedback stability analysis. As described earlier, plotting the 1/ß curve with the amplifier open-loop gain response permits the evaluation of feedback stability, and this analysis depends upon the frequency dependence of ß. For the present example the impedance characteristics of the load affect ß, and oscillation can result when an inductive load, such as a motor, replaces R_L. Then the load impedance rises with increasing frequency, causing a corresponding decrease in ß. This makes the 1/ß curve rise with frequency, and as discussed in Sec. 1.3.1, a rising 1/ß curve signals potential oscillation. To retain stability in these cases, a capacitance bypass of load R_L levels off the circuit's 1/ß curve.

2.4.4 Completing the analysis

Continuation of the analysis example yields the circuit's transfer response and dynamic range limit. The redrawn circuit of Fig. 2.13

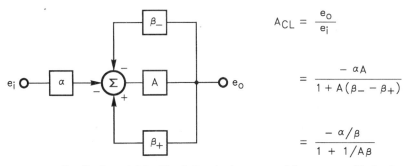

Figure 2.14 Feedback model for Fig. 2.13 substitutes α and ß attenuator blocks for voltage divider actions of the circuit.

translates to the feedback model of Fig. 2.14. In the model, the $ß_-$ and $ß_+$ blocks represent the voltage dividers of the circuit's negative and positive feedback networks. Also, the α block represents the attenuation applied to e_i by the negative feedback network. Signals from the three attenuator blocks connect to the summation element of the model, and there the summation polarities correspond to the op amp input polarities in the circuit. Finally, the summation output drives gain block A, and this block delivers the output signal e_o.

This feedback model predicts the amplifier output voltage e_o resulting from an input signal e_i, but the desired circuit response for the example remains the output current i_L. A mixture of model analysis and circuit analysis yields the desired result, and this mixture illustrates the feedback analysis approach for any op amp circuit where the op amp and the circuit outputs differ. For the present example, model analysis produces the e_o/e_i response and circuit analysis translates this to the i_L/e_i result. With the model, a loop equation defines the e_o/e_i response starting with $e_o = A(-\alpha e_i - ß_- e_o + ß_+ e_o)$. Solving this expression produces the $A_{CL} = e_o/e_i$ result, and further manipulation delivers the standard form with a $1 + 1/Aß$ denominator. Then, for Fig. 2.14,

$$A_{CL} = \frac{e_o}{e_i} = \frac{-\alpha/ß}{1 + 1/Aß}$$

Translation of this result to the desired i_L/e_i response requires a return to the circuit diagrams of the example. In Fig. 2.12, the output current produces the load voltage $e_L = i_L R_L$. In Fig. 2.13, the output voltage e_o transfers to load R_L, included in $R_2 || R_L$, through the positive feedback factor $ß_+$. There the load voltage equals the positive feedback signal of $e_L = ß_+ e_o$. Combining these two equations produces $e_o = i_L R_L/ß_+$, and this describes the op amp output voltage resulting from a given output current and load. This expression communicates the dynamic range limit imposed by the amplifier's output voltage range limit. Substitution of this expression in the preceding A_{CL} equation also yields the desired circuit response of

$$\frac{i_L}{e_i} = \frac{-1/R_2}{1 + 1/A\beta}$$

As before, the response numerator defines the ideal closed-loop gain, and the standard form denominator transfers general bandwidth and stability results to this response.

2.5 Generalized Feedback Model

Drawing and analyzing feedback models adds insight into op amp circuit operation, and this approach simplifies the circuit analysis for any op amp application. However, a generalized feedback model and standard response equations further simplify the analysis process. These standardized results avoid even the single-loop equation of the model analysis by expressing the circuit response through just the α and ß factors. Op amp circuit analysis then reduces to voltage divider analysis, and simple inspection typically defines the associated divider ratios. The generalized model and its performance equations apply to all op amp circuits.

2.5.1 Generalized model and its performance results

Figure 2.15 shows the generalized feedback model and the associated response equations. Connections to the model's summation element include all of the four possible input and feedback connections to the two inputs of an op amp. Each model connection transfers the signal to the summation element through an α or ß attenuation block, but only some of these connections apply in most op amp configurations. In those cases, the other α or ß terms of the model simply become zero. Similarly, many op amp applications make direct input or feedback connections to the op amp inputs. No attenuation affects these direct connections, and the associated α and ß terms equal unity. For the example of Fig. 2.13, the circuit lacks input signal coupling to the op amp's noninverting input. This sets $\alpha_+ = 0$ and the model in Fig. 2.15 reduces to that of Fig. 2.14. Similarly, the circuit in Fig. 2.10 connects the input signal directly to the amplifier's noninverting input, and this amplifier input lacks a feedback connection. Then $\alpha_+ = 1$ and $\beta_+ = 0$ and the generalized model reduces to that of Fig. 2.11.

Analysis of the generalized model yields standardized equations

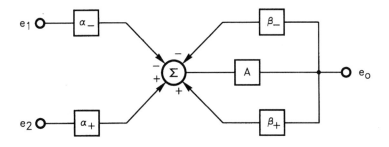

$$A_{CL} = \frac{e_o}{e_i} = \frac{\alpha/\beta}{1 + 1/A\beta} = \frac{A_{CLi}}{1 + 1/A\beta}$$

e_i	α
e_1	$-\alpha_-$
e_2	α_+
$e_2 - e_1$	$\alpha_+ = \alpha_-$
$e_1 = e_2$	$\alpha_+ - \alpha_-$

$\beta = \beta_- - \beta_+$

$$A_{CLe} = \frac{1/\beta}{1 + 1/A\beta} \qquad BW = \beta f_c$$

Figure 2.15 Generalized feedback model, with standardized response equations, reduces op amp circuit analysis to determination of α and β voltage-divider ratios.

that also adapt to any specific op amp application. For the model, analysis defines the closed-loop response as

$$A_{CL} = \frac{e_o}{e_i} = \frac{\alpha/\beta}{1 + 1/A\beta} = \frac{A_{CLi}}{1 + 1/A\beta}$$

where $\beta = \beta_- - \beta_+$. Here α and e_i depend upon the actual circuit configuration, but the form of the response remains the same for all cases. The next section defines the specific α and e_i terms for the general op amp configurations.

The form of the A_{CL} equation links the generalized model to feedback performance results of earlier sections of this chapter and of Chap. 1. Reducing the A_{CL} denominator to the form $1 + 1/A\beta$ links the earlier results through both the equation's denominator and its numerator. First this standard denominator transfers the Chap. 1 closed-loop bandwidth and frequency stability conclusions to the model here. Reducing the A_{CL} equation to express this standard denominator conclusively identifies ß, defining the closed-loop bandwidth through BW = ßf_c. Section 1.2 develops this bandwidth relationship by demonstrating the coincidence of the bandwidth limit with the intercept of the 1/ß and A curves. Also at this intercept, the curve slopes predict frequency stability, as described in Sec. 1.3. This stability analysis requires knowledge of the net circuit ß. Solution for the standard form denominator defines ß = ß$_-$ − ß$_+$ for the model, and this matches the multiple feedback result of Sec. 2.2 by modeling the differential input action of op amps.

Also, repetition of the 1 + 1/Aß denominator links the model to a second bandwidth limit described previously for the error signal gain. This gain amplifies the differential input error signal of an op amp, reflecting all major amplifier error effects to the circuit output. At low frequencies, this gain equals 1/ß, as described in Sec. 1.1.2. At higher frequencies, the bandwidth limit contained in this standard denominator also applies to the error signal gain, as described in Sec. 1.2.4. There the basic error gain of 1/ß responds to the same high-frequency roll off as A_{CL}. Combining the initial 1/ß gain with this denominator repeats the equation developed for error signal gain in Sec. 1.2.4. All op amp circuits amplify the input-referred errors of Sec. 1.1 by a gain of

$$A_{CLe} = \frac{1/ß}{1 + 1/Aß}$$

2.5.2 Applying the generalized result

So far previous performance results apply directly to any circuit analyzed with the generalized feedback model. However, application of the generalized A_{CL} equation still requires the determination of the net α for a given circuit's configuration. Different input connections result in different α factors, and the determination of a circuit's net α concludes the solution for the configuration's ideal closed-loop gain. The preceding generalized A_{CL} expression communicates this ideal gain through its numerator. In the ideal case, high loop gain Aß reduces the denominator of this expression to unity, leaving the closed-loop gain A_{CLi} = α/ß. For this result, previous discussion shows the net ß = ß$_-$ − ß$_+$ to be the same for all configurations. However, the net α varies with the circuit configuration.

Fortunately, all practical op amp configurations produce only four possible α results. configuration differences affect the net circuit α through varied input connections to the α$_-$ and α$_+$ blocks of the generalized model. For that model, a given combination of the e_1 and e_2 signals defines a net input signal e_i and determines the participation of α_- and α_+. Different conditions for e_1 and e_2 represent the inverting, noninverting, differential, and common-mode input configurations of op amps.

Model analysis of each configuration defines the corresponding net α and the relevant e_i. Comparison of each analysis result with the generalized A_{CL} expression then defines these parameters for each configuration. For the inverting configuration, setting e_2 to zero leaves only e_1 connected to the model, and this signal connects through the α_- block. Under these conditions, analysis of the model defines net circuit parameters of $\alpha = -\alpha_-$ and $e_i = e_1$. Similar analysis for the noninverting case sets e_1 to zero and produces $\alpha = \alpha_+$ and $e_i = e_2$. Differential input connections apply to both e_1 and e_2 and impose the practical requirement that $\alpha_+ = \alpha_-$. This requirement balances the circuit to produce the subtraction of the circuit's differential action. Then model analysis yields $\alpha = \alpha_+ = \alpha_-$ and $e_i = e_2 - e_1$. Finally the common-mode case requires that $e_i = e_2 = e_1$ and produces a net $\alpha = \alpha_+ - \alpha_-$. The following table summarizes the results.

Input connection	e_i	α
Inverting	e_1	$-\alpha_-$
Noninverting	e_2	α_+
Differential	$e_2 - e_1$	$\alpha_+ = \alpha_-$
Common mode	$e_1 = e_2$	$\alpha_+ - \alpha_-$

Examination of the table shows agreement with previous modeling results. From the table an inverting input connection with $e_i = e_1$ corresponds to $\alpha = -\alpha_-$. Previously, Fig. 2.2b demonstrated this result, where the $-\alpha$ in the A_{CL} response numerator there represents $-\alpha_-$ here. Similarly, Fig. 2.8 supports the noninverting table entry with a numerator term of $\alpha = \alpha_+$. Finally, Fig. 2.11 shows a common-mode input case producing a $1-\alpha$ factor in the response numerator, and this factor corresponds to the $\alpha_+ - \alpha_-$ of the table, where $\alpha_+ = 1$.

2.6 Analysis of Complex Feedback

All of the preceding analysis examples address fairly simple feedback configurations, which include a single op amp, a fixed feedback factor, and only one feedback connection per amplifier input. However, op amp circuits sometimes employ more complex connections. Some include additional amplifiers or other active elements in the feedback path. Depending upon their placement, these added elements modify either the feedback factor or the open-loop gain, or both. Some of these added feedback elements also produce variable feedback factors that define a range of operation for the feedback analysis. Other circuits include multiple input or multiple feedback connections at a given amplifier input. In these multiple-connection cases, analysis shows that the signal effects simply add up.

The generalized feedback model does not directly represent the added elements of these complex circuits. As described in the preceding section, this model includes a gain block for one gain element and one α and ß block per amplifier input. Thus the standardized results of the model do not immediately apply to the more complex circuits. However, simple analysis adapts the more complex circuit conditions to the model blocks. Using the feedback principles of Sec. 2.4, analysis first defines net results for the open-loop gain, $ß_-$ and $ß_+$ for a given circuit. This consolidates the effects of multiple gain elements and multiple feedback connections. Then the net A, $ß_-$, and $ß_+$ of the circuit simply become the A, $ß_-$, and $ß_+$ of the generalized model.

Next, superposition analysis separates the different α terms that result from multiple signal connections at a given amplifier input. Such connections produce multiple α_- or α_+ factors. Each factor corresponds to only one input signal, requiring a separate analysis for each input case. Superposition analysis identifies the various α terms for separate application to the feedback model. However, the generalized model only accommodates one input connection to each amplifier input through the α_- and α_+ blocks. To accommodate multiple connections, a different α_- or α_+ defines the model block for each case.

This process reduces any complex op amp feedback connection to the level of the generalized feedback model, again simplifying circuit analysis. Then the model represents the circuit for a specific case, transferring standard results to the circuit analysis. Previously for simple feedback, this transfer reduced op amp circuit analysis to the determination of voltage divider ratios. To this process, complex feedback simply adds a multiplication of the gain and divider effects and a separation analysis cases for different input signals.

2.6.1 Multiple amplifiers and the ß feedback block

An added amplifier in the feedback loop modifies either the net feedback factor or the net open-loop gain, or both. Figure 2.16 illustrates the modified feedback factor case with a specialized difference amplifier configuration that changes the circuit's gain without disturbing common-mode rejection. As shown, equal-valued R resistors set the gain of the basic difference amplifier at unity. For other gains, the ratios of the R resistors would normally be changed. However, for

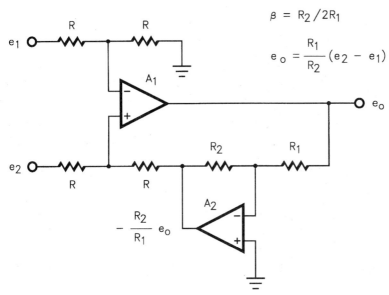

Figure 2.16 Addition of the A_2 inverter changes the feedback factor of a difference amplifier to alter closed-loop gain without disturbing common-mode rejection.

committed difference amplifiers, very precise manufacturing adjustment has already set these ratios for high common-mode rejection.

To retain the benefit of this adjustment, the circuit shown changes gain through the addition of amplifier A_2. Analysis of this circuit could be performed using a feedback model having A, α, β, and Σ elements for each amplifier. Instead, more direct analysis includes the effect of A_2 in a net feedback factor for the circuit. The position of A_2 in the circuit corresponds to the β_+ path of the generalized model of Fig. 2.15, and including the effect of A_2 in β_+ makes that model and its standardized results apply to this case. The inverting amplifier formed with A_2 inverts and scales the output signal that drives the difference amplifier feedback. Then the fraction of the output e_o fed back to the A_1 noninverting input becomes $\beta_+ = -(\frac{1}{2})R_2/R_1$. No feedback drives the A_1 inverting input so $\beta_- = 0$, and this makes the net feedback factor $\beta = \beta_- - \beta_+ = R_2/2R_1$.

Using this net feedback factor, the application of the generalized model equations defines the circuit response. For that model, $A_{CLi} = e_o/e_i = \alpha/\beta$ represents the ideal closed-loop gain. Also, for the differential input case considered here, the standardized results define $e_i = e_2 - e_1$ and $\alpha = \alpha_1 = \alpha_2$. The equal-valued resistors of the difference amplifier shown make $\alpha_1 = \alpha_2 = \frac{1}{2}$. Substitution of the e_i, α, and β results in the A_{CLi} equation leads to the ideal circuit response of

$$e_o = \frac{R_1}{R_2}(e_2 - e_1)$$

Thus the circuit provides the desired change from the unity gain of the original difference amplifier.

Bandwidth and stability analyses for the circuit also benefit from previous standard results, and Fig. 2.17 shows the relevant response curves. There the open-loop gain curve of A_1 and the 1/ß curve corresponding to the net feedback factor of the circuit define performance. The 1/ß curve includes the effects of A_2 for this analysis. As illustrated, the two curves intersect at f_i with a comfortable 20-dB per decade rate of closure. From before, this intercept predicts good frequency stability and a circuit bandwidth of BW = $\text{ß}f_{c1}$. However, this circuit's stability can be compromised by the bandwidth limit of A_2. Note that the 1/ß curve rises beginning at $\text{ß}_2 f_{c2}$ due to the bandwidth limit of A_2. Here $\text{ß}_2 = R_1/(R_1 + R_2)$ represents the feedback factor of the A_2 inverter, and f_{c2} is the unity-gain crossover frequency of A_2. As long as $\text{ß}_2 f_{c2} \geq 10 f_i$, the 1/ß rise produces little stability disturbance. However, for other cases the stability analysis methods of Sec. 1.3 define the result and determine the need for corrective phase compensation.

2.6.2 Multiple amplifiers and the A gain block

Other feedback analysis cases include the effect of an added amplifier in a net open-loop gain. For example, the composite amplifier of Fig.

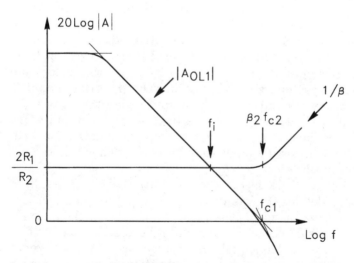

Figure 2.17 Bandwidth and stability analysis for Fig. 2.16 employs the open-loop gain of A_1 and the net circuit feedback factor, as altered by A_2.

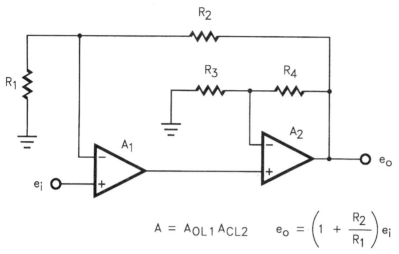

Figure 2.18 A composite amplifier increases net open-loop gain by adding a second amplifier in series with the gain path.

2.18 introduces a second amplifier between the circuit's input and output. As will be seen, the gain of the second amplifier reduces the composite circuit's gain error and increases its gain-bandwidth product. This gain also increases the net open-loop gain enclosed by the overall feedback network of the circuit. Resistors R_1 and R_2 supply the overall feedback whereas R_3 and R_4 provide a localized feedback for amplifier A_2.

In this composite amplifier, the gain supplied by the second op amp supplements the gain block path of the generalized model in Fig. 2.15. Including the A_2 effect in the A block transfers the generalized results to this example. This composite amplifier connects the two amplifiers in series within a common feedback loop to produce a net open-loop gain of $A = A_{OL1}A_{CL2}$. Here A_{OL1} is the open-loop gain of amplifier A_{OL1} and $A_{CL2} = 1 + R_4/R_3$ is the closed-loop gain set by the separate A_2 feedback. Simple application of this net gain with the generalized model and its standardized equations defines the composite circuit response.

Graphical analysis best demonstrates the results, as shown in Fig. 2.19. There the net gain curve A results from its components A_{OL1} and A_{CL2}. On the logarithmic gain scales of the graph, linear summation of the component curves produces the net gain curve. This summation transfers the closed-loop response pole of A_2 to the net gain, producing a two-pole response. The second pole occurs at a frequency of $\beta_2 f_{c2}$ and generally restricts the use of the composite amplifier to higher closed-loop gains. Such gains place the 1/ß intercept in a region of

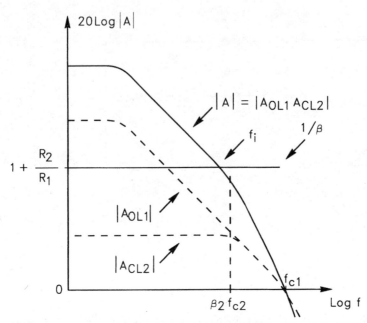

Figure 2.19 A_{CL2} gain added by the second amplifier of Fig. 2.18 shifts the open-loop response upward to increase both loop gain and circuit bandwidth.

reduced gain slope as illustrated. For lower-gain applications, phase compensation addresses other solutions to the two-pole response slope.[4,5] However, for the example shown, the overall circuit feedback defines a constant $1/ß = 1 + R_2/R_1$ for the flat response curve shown. Amplifier A_2 does not affect this curve because the net gain curve A includes the effect of this amplifier.

These curves show improved circuit accuracy and bandwidth as a result of the composite amplifier configuration. The increased loop gain reduces the gain error signal e_o/A described in Sec. 1.1.2. This error signal develops between the amplifier inputs, resulting in an output error of $e_o/Aß$, where the error-reducing $Aß$ term represents the circuit's loop gain. As described in Sec. 1.2.2, $Aß$ corresponds to the vertical distance between the A and $1/ß$ curves. For the composite amplifier, this distance increases as A_{CL2} moves the circuit gain from A_{OL1} to the net circuit gain A in the figure. Similarly, bandwidth increases as the f_i intercept moves from the intersection with the A_{OL1} curve to that with the net gain $A = A_{OL1}A_{CL2}$ curve. Separate analysis shows the bandwidth to be BW = $A_{CL2}ßf_{c1}$, so the composite amplifier increases the bandwidth by a factor equal to A_{CL2}.

2.6.3 Variable feedback factors

All of the preceding examples of this chapter have fixed feedback factors, but some op amp applications produce variable feedback factors through potentiometers, switches, or analog multipliers. Then both the magnitude and the frequency characteristics of 1/ß can become variables. Still, the determination of a net circuit feedback factor or a net closed-loop gain adapts these circuits to the generalized results. This permits simple performance evaluation over the range of conditions presented by the varying feedback conditions.

Magnitude variation in the feedback factor results with a common analog divider circuit. There placing a multiplier in the feedback loop of an op amp[6,7] inverts the multiplier function to produce divider operation. Shown in Fig. 2.20, this circuit ideally produces a divider response of $e_o = -10R_2e_i/R_1e_c$. Feedback analysis determines the actual circuit performance and the relative frequency stability of the circuit. This analysis could follow the modeling process of Sec. 2.4 with the multiplier represented by a separate block in the feedback path. However, including the multiplier effect in ß permits use of the generalized feedback model and that model's standardized results. The multiplier response scales the fraction of the amplifier output fed back to the input; otherwise the circuit remains a simple inverting amplifier, easily represented by the generalized model. Without the multiplier, the basic inverting amplifier supplies the output signal e_o directly to R_2 and produces a feedback factor of $R_1/(R_1 + R_2)$. Including the multiplier, this device's transfer function of $XY/10$

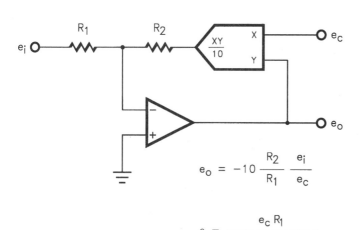

Figure 2.20 A multiplier in an op amp feedback loop produces divider operation through a variable feedback factor.

scales the signal supplied from e_o to a level of $e_o(e_c/10)$. The same scaling applies to the feedback factor, and for the divider circuit shown,

$$\beta = \frac{e_c R_1}{10(R_1 + R_2)}$$

This defines the circuit's net feedback factor as a function of a control signal e_c.

Using this expression for the circuit's feedback factor extends the previous feedback analysis results to the divider circuit. With a signal-dependent feedback factor, the bandwidth and stability conditions become variables, as shown in Fig. 2.21. There variation of the control signal e_c moves the 1/ß curve across the full range of the op amp's gain-magnitude response. As e_c nears zero, the 1/ß curve approaches infinity, leaving the op amp in essentially an open-loop condition. At the other extreme, a value of $e_c = 10$ V delivers the full e_o signal to R_2 as if the multiplier were replaced by a short circuit. Then the circuit acts as a simple inverting amplifier with a feedback factor of $R_1/(R_1 + R_2)$ and an inverting gain of $-R_2/R_1$.

Between these extremes, the variation of e_c moves the 1/ß curve from as low as the unity-gain axis to above the upper reaches of the op amp's gain magnitude curve. This moves the critical intercept and requires attention to its rate of closure over the entire gain response. Note that this rate of closure also depends upon the bandwidth of the multiplier, and to avoid stability degradation, the multiplier band-

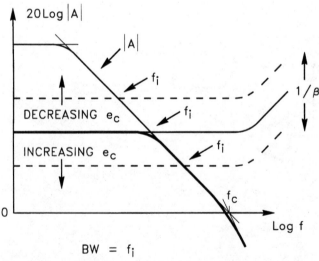

Figure 2.21 Variation in multiplier control voltage e_c of Fig. 2.20 moves the 1/ß curve up and down the amplifier open-loop response, producing bandwidth variations reflected by f_i intercepts.

width must be much greater than that of the op amp. Then the 1/ß curve remains flat at the moving intercept, and just a unity-gain stable op amp assures frequency stability for the overall circuit. The moving intercept described defines the range of bandwidth for the divider operation as indicated by the f_i frequencies shown. From before, BW = f_i = ßf_c. For a given e_c range, the intercept moves linearly with e_c, defining a corresponding range for bandwidth.

A more restricted multiplier bandwidth compromises frequency stability. Poles in the multiplier response produce a roll off of the feedback factor that makes the circuit's 1/ß curve rise at higher frequencies. Figure 2.21 illustrates this effect for a wide-band multiplier case that keeps this rise above the f_i intercept. However, the 1/ß rise draws toward the critical intercept when the multiplier control voltage e_c increases and moves the 1/ß curve downward in the graph. This moves the 1/ß rise closer to the resulting f_i intercept, potentially compromising the circuit's rate of closure there. To avoid instability, the circuit's op amp must introduce a dominant pole at a low enough frequency to limit that rate of closure.

2.6.4 Multiple input connections

The preceding feedback examples illustrate only one connection to a given op amp input. Other op amp applications make multiple connections to an input, as in the simple summing amplifier of Fig. 2.22. There three example input signals connect to the op amp's inverting input through different summing resistors. Thus different feedforward α factors apply for each input signal. However, the generalized

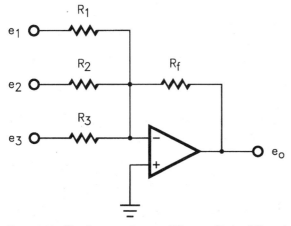

Figure 2.22 Simple summing amplifier results in different feedback conditions for each input signal.

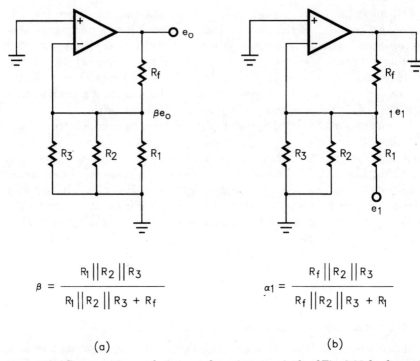

Figure 2.23 Superposition analysis grounds various terminals of Fig. 2.22 for determination of ß and α terms.

feedback model includes only one α block for representing the signal connection to the inverting amplifier input.

Superposition analysis adapts this summing amplifier to the model by considering one input signal at a time. In the process, the analysis demonstrates that multiple input connections having different α terms still have the same feedback factor ß. Figure 2.23 shows the summing amplifier redrawn for superposition analysis of both ß and α. Superposition grounds all but one signal source and analyzes the effect of that remaining source. To determine first ß, superposition sets all of the input signals to zero, leaving e_o as the only nonzero signal, as shown in Fig. 2.23a. This grounds the input sides of R_1 through R_3, placing these resistors in parallel and making the feedback factor for each input case of the example

$$\beta = \frac{R_1 || R_2 || R_3}{R_1 || R_2 || R_3 + R_f}$$

However, the superposition grounding produces different effects for the α attenuator terms. Consider the attenuation received by the input signal e_1, as illustrated in Fig. 2.23b. Superposition then zeros or grounds e_2, e_3, and e_o. This places R_2, R_3, and R_f in parallel and connected to ground. Then the voltage divider ratio defining α_1 is simply

$$\alpha_1 = \frac{R_f||R_2||R_3}{R_f||R_2||R_3 + R_1}$$

Analogous divider ratios define α_2 and α_3, and they differ only in the parallel resistor connections produced by the superposition grounding. For α_2, simply interchange R_1 and R_2 in the preceding expression, and this produces

$$\alpha_2 = \frac{R_f||R_1||R_3}{R_f||R_1||R_3 + R_2}$$

Similarly, for α_3, interchange R_1 and R_3 in the α_1 expression. This process extends to any number of signal input connections beyond the three shown in the example.

The determination of the ß and α factors of the summing amplifier transfers all of the previous generalized results to this multiple-input case. Bandwidth and frequency stability remain a function of the $1 + 1/Aß$ response of the denominator. The same ß applies for all inputs, so these performance characteristics remain the same for all input α connections. Bandwidth remains BW = $ßf_c$ for all input signals, regardless of the closed-loop gain received by a given signal. Note that each added input connection reduces ß and this explains the restricted bandwidth often encountered with summing amplifiers.

The α differences of multiple inputs reflect in different closed-loop gains and the generalized model reflects this in the ideal closed-loop gain $A_{CLi} = \alpha/ß$. Again using superposition, the α term corresponding to a given input independently defines the associated ideal gain. For the various inputs, this expression identifies different closed-loop gains at lower frequencies, but all of these gains roll off at the same frequency. Fundamental to op amp circuits, response roll off defines a bandwidth limit at BW = $ßf_c$ independent of the signal input considered with the summing amplifier.

2.6.5 Multiple feedback paths

Other op amp configurations make multiple connections to an op amp input through different feedback paths. The resulting multiple feedback factors add directly, producing a net feedback factor for the circuit. In this addition, differing feedback gains scale the relative contributions of the individual feedback factors. However, a determination of the net feedback factor again extends the generalized feedback model and its standardized results to this multiple-feedback case.

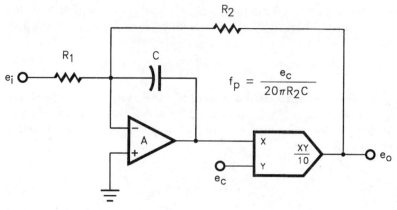

Figure 2.24 A voltage-controlled low-pass filter introduces a second feedback path to the amplifier inverting input through attenuated signal drive.

Figure 2.24 shows a multiple-feedback example that also illustrates the multiple gain elements described earlier. In the figure, feedback to the amplifier inverting input originates from both the amplifier output and the output of an added multiplier. The multiplier acts as a gain element in series with the input-to-output path of the circuit, altering the net circuit gain as well as the feedback factor. In this way, the multiplier and its resistor feedback connection convert the circuit from a simple integrator to a voltage-controlled low-pass filter. Replacing the multiplier with a short circuit illustrates the basic low-pass filter. This short-circuited condition essentially results when $e_c = 10$ V sets the gain through the multiplier at unity. Then the op amp, the resistors, and the capacitor form an inverting amplifier with feedback bypass. Bypass capacitor C then breaks with R_2 to define the filter roll off, just as if the resistor and the capacitor were directly in parallel. For this short-circuit condition, the resulting break frequency occurs at $f_p = 1/2\pi R_2 C$.

For levels of e_c below 10 V the multiplier serves as a voltage-controlled attenuator to effectively alter the filter time constant. The typical $XY/10$ transfer response of the multiplier sets the attenuation factor for the X input signal at $e_c/10$. Once attenuated, the signal drives the R_2 feedback resistor. Attenuating the feedback voltage delivered to R_2 reduces the signal current delivered to the op amp's summing node. This produces the same result as if the resistor had been increased in value to $10R_2/e_c$, and this result reflects an effective resistance for the circuit feedback. Increased effective resistance corresponds to a decreased break frequency with the capacitor C. Substituting $10R_2/e_c$ for R_2 in the f_p expression of the preceding paragraph defines the circuit's variable break frequency as

$$f_p = e_c/20\pi R_2 C$$

The variable feedback attenuation described moves the corresponding $1/\beta$ curve, requiring performance analysis over a range of conditions. This analysis reduces the circuit feedback characteristics to net circuit terms for application of the generalized feedback model. As described in Sec. 2.5.1, this model produces a transfer response of

$$A_{CL} = \frac{e_o}{e_i} = \frac{\alpha/\beta}{1 + 1/A\beta} = \frac{A_{CLi}}{1 + 1/A\beta}$$

Expressing the transfer response in Fig. 2.24 in the same form extends the previous results for the error, bandwidth, and stability analysis to this new circuit. However, this requires determination of the circuit's A_{CLi}, A, and β terms.

Simple inspection of the circuit defines the corresponding response numerator A_{CLi}, and from before, this ideal gain has a pole at f_p. Prior to this roll off, the feedback capacitor C does not affect the circuit at the low frequencies of the pass band, leaving R_1 and R_2 to control the circuit with simple inverting amplifier feedback. This makes the pass-band gain $-R_2/R_1$. Combining this pass-band gain with the pole at f_p produces a single-pole low-pass filter having an ideal response of

$$A_{CLi} = \frac{-R_2/R_1}{1 + (10R_2/e_c)Cs}$$

Further analysis defines the actual A_{CL} response denominator through the net β and A terms of the circuit. First the amplifier and multiplier set the net input-to-output open-loop gain as described in Sec. 2.6.2. The amplifier contributes an open-loop gain of A_{OL1} and the multiplier inserts an attenuation of $e_c/10$ for a net open-loop gain of $A = A_{OL1}e_c/10$. The net β for the circuit depends upon the effects of the two feedback paths, and the application of the feedback analysis process described in Sec. 2.4 quantifies the corresponding feedback effects. Following that process, Fig. 2.25 displays the feedback model for the circuit of Fig. 2.24. The model represents the multiplier by a gain block having that element's typical transfer response $XY/10 = Xe_c/10$. The β_C and β_R blocks represent the two circuit feedback connections through the capacitor and the resistor. Both feedback blocks connect to negative inputs at the model's summation element in correspondence with the inverting input connections of the circuit. Finally, an input α attenuator block represents the input signal transmission to the amplifier's summing node.

Analysis of this model shows that the effects of multiple feedback connections add directly to define the net feedback factor for the circuit. Writing and manipulating the model's input-to-output transfer

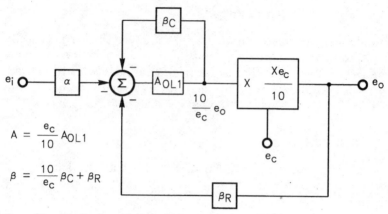

Figure 2.25 Feedback modeling of Fig. 2.24 reveals circuit net gain and feedback factor, linking the circuit with the generalized feedback model.

response e_o/e_i, displays the net ß. Further manipulation of the response expression into the standard form of the A_{CL} expression reduces the denominator to the form $1 + 1/Aß$. From before, the net gain is $A = A_1 e_c/10$. Including this result in the analysis shows the net feedback factor for Fig. 2.24 to be

$$ß = (10/e_c)ß_C + ß_R$$

Here the two individual feedback factors contribute to the net circuit ß in different ways. First the direct contribution of $ß_R$ reflects that feedback's direct connection between the circuit output and the amplifier input. However, the modified contribution of $ß_C$ reflects the intervening attenuation of the multiplier. The op amp output drives $ß_C$ prior to this attenuation, and there the signal level is $(10/e_c)e_o$. This increases the effect of the $ß_C$ feedback by the factor $10/e_c$, as expressed in the net ß equation. Thus multiple feedback to a given amplifier input produces a net feedback factor equal to a scaled sum of the individual factors. The scaling applied to each factor equals the ratio of the signal driving the corresponding feedback network to the circuit's output signal.

Direct mathematical analysis would define $ß_C$ and $ß_R$, yielding the net circuit ß for the example in Fig. 2.24. However, just the knowledge of the $ß_C$ and $ß_R$ contributions permits an intuitive analysis of the net effect. Figure 2.26 displays the resulting 1/ß and gain curves for stability analysis. As shown with dashed lines, open-loop gain A moves up and down with the variation of e_c in correspondence with the net circuit gain of $A = A_{OL1}e_c/10$. Other dashed lines also show that the 1/ß

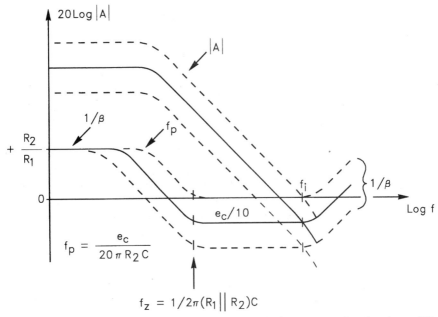

Figure 2.26 Control voltage e_c of Fig. 2.24 moves A and $1/\beta$ curves together, leaving stability conditions at intercept f_i unchanged

curve simultaneously changes with the variation in e_c, but in a more complex manner. Fortunately, the gain and $1/\beta$ curves move in unison, leaving the intercept at f_i unchanged. Thus the circuit's stability conditions remain unchanged as e_c varies the filter cutoff frequency.

To demonstrate this, consider the frequency regions where the two feedback paths dominate. At low frequencies, C remains an open circuit and the inverter feedback of R_1 and R_2 controls the circuit. There β_R dominates the net feedback factor and $1/\beta = 1/\beta_R = 1 + R_2/R_1$, as shown in the response curves. At higher frequencies, C becomes a short circuit and supplies unity feedback around the op amp. There $\beta_C = 1$ dominates and $1/\beta = (e_c/10)\beta_C = e_c/10$, and this remains the $1/\beta$ condition at the three f_i intercepts shown in Fig. 2.26. At these intercepts, $1/\beta = e_c/10$ and $A = A_{OL1} e_c/10$, so both curves move with the same dependence upon the control voltage e_c. The characteristics of the intercept remain unchanged, leaving stability unaffected by the variation in e_c.

The stability evaluation of this example illustrates two unusual feedback analysis conditions. The $1/\beta$ curve rises at higher frequencies and the intercept at f_i drops below the 0-dB or unity-gain axis. First the $1/\beta$ curve rises due to the roll off of the multiplier bandwidth, as described previously with Fig. 2.21. For good stability, this bandwidth limit must be well above the intercept frequency f_i. Next the intercept below the unity-gain axis results from the attenuation of the multiplier. With almost all other feedback connections, the 1/ß curve never drops below unity because this unity level corresponds to a maximum possible feedback factor of 1 that governs most circuits. There the maximum fraction of the amplifier output fed back to the input equals 100%, precluding any increase in this feedback fraction. However, the multiplier attenuation of this circuit results in a larger feedback signal driving β_C. This makes the fraction of e_o fed back to the input greater than 1, driving the 1/ß curve below the unity-gain axis. Still, stability analysis criteria remain the same, and the phase shift at the f_i intercept remains the defining condition.

To complete the 1/ß curve, consider the transition between the low-frequency and high-frequency regions of that curve. The transition from $1 + R_2/R_1$ to $e_c/10$ represents the transition in the control of the two feedback paths. This begins at the break frequency of C and the effective resistance of $10R_2/e_c$ or at $f_p = e_c/20\pi R_2 C$. Following f_p, the 1/ß curve rolls off with a single-pole 20-dB per decade slope down to the $e_c/10$ level. There a zero at $f_z = 1/2\pi(R_1 || R_2)C$ levels off the 1/ß curve. Note that f_z represents the break frequency of the capacitance and the net resistance connected to the summing node. From that node, both R_1 and R_2 connect to low impedances, making the net resistance presented to the node $R_1 || R_2$.

References

1. J. Graeme, "Feedback Models Reduce Op Amp Circuits to Voltage Dividers," *EDN*, June 20, 1991, p. 139.
2. J. Graeme, "Generalized Op Amp Model Simplifies Analysis of Complex Feedback Schemes," *EDN*, April 15, 1993, p. 175.
3. M. Stitt and R. Burt, "Möglichkeiten zur Rauschunterdrückung," *Elektronik*, December 1987.
4. J. Graeme, "Composite Amplifier Hikes Precision and Speed," *Electron. Des.*, June 24, 1993, p. 64.
5. J. Graeme, "Phase Compensation Perks up Composite Amplifiers," *Electron. Des.*, August 19, 1993, p. 30.
6. Y. Wong and W. Ott, *Function Circuits: Design and Applications*, McGraw-Hill, New York, 1976.
7. J. Graeme, "Bode Plots Enhance Feedback Analysis of Operational Amplifiers," *EDN*, February 2, 1989, p. 163.

Chapter

3

Power-Supply Bypass

Habit rather than analysis often erroneously guides the mundane selection of power-supply bypass. This practice suboptimizes the rejection of power-supply noise and can result in oscillation, especially with higher-frequency op amps. Analysis of these noise coupling and stability problems reveals a complex relationship between an op amp and its supply-line impedances. There inductive effects and multiple impedance resonances compromise the simple capacitance shunting desired from the bypass. Optimizing the bypass result yields simple design equations that provide a more guided approach to the bypass selection. This chapter first explores the nature of power-supply noise coupling and the multiple effects of a single bypass capacitor.[1] Then discussions of dual bypass capacitors, bypass detuning, and power-supply decoupling expand the bypass alternatives.[5]

3.1 Power-Supply Bypass Requirement

The need for power-supply bypass arises from parasitic supply-line impedances that degrade noise and stability performance. Line impedance interaction with currents drawn by the amplifier, and other circuitry powered from the same supply lines, produces supply-line noise signals. There supply current drains flow through this impedance, producing voltage drops that represent noise at the amplifier supply connections. This supply noise reacts with the amplifier's power-supply rejection ratio (PSRR), reproducing a portion of the noise signal at the op amp inputs. There the coupled noise combines with that of the amplifier to be amplified by the noise gain of the application circuit. In this coupling mechanism the portion of the supply noise resulting from the amplifier itself constitutes a parasitic feedback signal capable of producing oscillation.

Numerous techniques commonly reduce this parasitic coupling through wide supply bus traces, minimized bus lengths, star connections, and supply bypass. The first two techniques, bus width and length control, reduce supply-line resistance and inductance. Star connections separate the supply runs serving sensitive and noisy circuitry, such as analog and digital, minimizing device interaction. However, most importantly, capacitive bypass of the power-supply lines attenuates the noise interference and ensures frequency stability as well. To be successful, the bypass selection requires close attention to the frequency-dependent impedances of both the supply lines and the bypass capacitors. Initial intuition suggests simply placing the bypass capacitors right at the amplifier's power-supply terminals and making the capacitors large.

Here the first intuitive assumption works, but the second fails. Close placement minimizes the interconnect length and the associated inductance between the amplifier and the capacitors. For best results, the bypass capacitors should be surface-mount types placed right at the amplifier's power-supply pins and returned through a ground plane. Practical layout considerations often preclude this ideal condition and, in any case, amplifier lead inductance remains unbypassed. The net, unbypassed inductance presents a fundamental limit to bypass effectiveness. The second intuitive assumption, making the capacitors large, fails because of the parasitic impedances of such capacitors. Large capacitor values do minimize the bypassed line impedances at lower frequencies where the capacitors remain purely capacitive. However, the internal construction of larger capacitors produces a significant parasitic inductance, interrupting the bypass effectiveness at higher frequencies. In compromise, the bypass capacitance value should be made large enough to sufficiently limit line impedance but no larger.

3.1.1 Noise coupling mechanism

Figure 3.1 illustrates the fundamental need for power-supply bypass. There positive and negative power supplies V_+ and V_- supply the corresponding bias terminals of the amplifier with the voltages V_P and V_N. Ideally, $V_P = V_+$ and $V_N = V_-$, but intervening supply-line inductances L_p react with the signal current i_S, producing voltage differences. Here the current i_S results from the combined supply currents drawn by the op amp and other circuitry powered from the same supply lines. The L_p inductances result from inherent parasitics of the connecting lines between the power supplies and the amplifier.

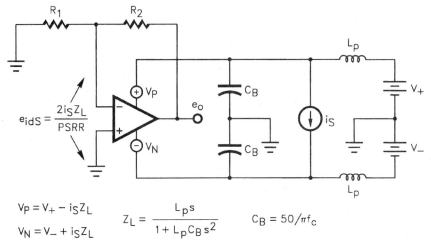

Figure 3.1 C_B bypass capacitors suppress supply voltage changes produced by reaction of supply current drain i_S with power-supply-line inductances L_p.

Typically, a wire or printed-circuit board trace introduces around 15 nH of parasitic inductance per inch of length.[2] Complex wiring and board connections produce inductance in the hundreds of nanohenrys. While seemingly small, the resulting inductive impedances can seriously degrade noise performance and frequency stability. Adding the C_B bypass capacitors shown rolls off these impedances to attenuate the supply coupling effects.

The circuit model shown intentionally oversimplifies the actual conditions of a practical circuit to provide more intuitive insight into the bypass requirement. First the model omits the input signal to permit focus upon those signals relating to the line impedance and its bypass. Next the model simplifies the line conditions. In practice, the V_+ and V_- supplies add output impedances and noise sources of their own, different circuits draw currents at different points upon the supply lines, and the supply-line resistance adds to the voltage drops produced with i_S. However, the bypass practices developed with this simplified model also contend with these other effects.

Intuitive evaluation of the model provides a first insight into the bypass benefit. Fundamentally, the C_B bypass capacitors shunt the line impedances to reduce the supply-line voltage drops produced by i_S. However, from another perspective, the capacitors serve as local reservoirs for the immediate supply of higher-frequency current demands. Otherwise such currents encounter significant time delays in their travel from the power supplies and through the line inductances to the op amp. This delay produces phase shift in the associated amplifier response, and for a given time delay, the corresponding phase shift increases with increasing frequency. To counteract this, the charge stored on the bypass capacitors locally supplies much of the high-frequency current demand, decreasing the time delay and the corresponding phase shift.

Circuit analysis quantifies the bypass benefit. The bypass capacitors reduce but do not eliminate the power-supply coupling effects. Voltage differences still develop with the residual impedances of the L_p, C_B combinations. The net line impedance Z_L reacts with the signal current i_S, reducing the voltage magnitudes at both the V_P and the V_N terminals by the amount $i_S Z_L$. Then $V_P = V_+ - i_S Z_L$ and $V_N = V_- + i_S Z_L$, so the total supply voltage delivered to the op amp, $V_P - V_N = V_+ - V_- - 2i_S Z_L$, decreases by the amount $2i_S Z_L$. This decrease reacts with the PSRR of the op amp, producing an amplifier input error of $e_{idS} = 2i_S Z_L/\text{PSRR}$. The circuit amplifies this error signal with the same A_{ne} gain that amplifies the op amp's input noise voltage, producing an output noise component,

$$e_{noS} = \frac{2A_{ne} i_S Z_L}{\text{PSRR}}$$

For an op amp circuit, the noise gain A_{ne} equals $A/(1 + A\beta)$, where A is the amplifier's open-loop gain and ß is the circuit's feedback factor. For the generalized configuration of Fig. 3.1, $A_{ne} \cong (1 + R_2/R_1)/(1 + f/f_c)$, where f_c is the unity-gain crossover frequency of the op amp.[3]

3.1.2 Frequency response of supply noise coupling

Examination of the preceding expression reveals the amplifier's power-supply sensitivity and related frequency dependencies. The supply coupled noise e_{noS} depends upon the frequency variations of both the line impedance Z_L and the amplifier PSRR. Both characteristics can vary in manners that increase e_{noS} as the frequency increases. Comparison of unbypassed and bypassed cases reveals that supply bypass reduces the resulting e_{noS} response from a double-zero to the flat response normally expected for op amp noise. Unbypassed, a declining PSRR and a rising line impedance both introduce zeros in the e_{noS} response. Bypassed, a declining PSRR and a declining line impedance produce canceling effects and a flat frequency response. Examination of these effects adds insight into potential noise origins for troubleshooting a circuit, especially in light of the practical bypass deficiencies described later.

One e_{noS} response zero always results from the declining PSRR of the op amp. This decline increases the $e_{idS} = 2i_S Z_L/\text{PSRR}$ input error signal to in turn increase $e_{noS} = A_{ne} e_{idS}$. An approximation to the PSRR fre-

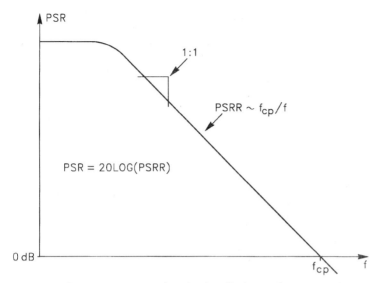

Figure 3.2 Op amp power-supply rejection displays a frequency response approximated by PSRR · f_{cp}/f, where f_{cp} is the unity-gain crossover of PSRR.

quency response simplifies the mathematical expression of this effect. Over most of the amplifier's useful frequency range, PSRR typically rolls off with a single-pole response, as shown in Fig. 3.2. In this single-pole span, the response drops with a 1:1 slope, making the PSRR decline in direct proportion to the increasing frequency. This declining curve crosses the unity-gain or 0-dB axis at a crossover frequency f_{cp}, marking a reference point. Together, f_{cp} and the curve's 1:1 slope define PSRR $\approx f_{cp}/f = \omega_{cp}/s$ in this dominant single-pole range. Substituting this PSRR approximation in the preceding e_{noS} equation produces $e_{noS} = 2A_{ne}i_S Z_L s/\omega_{cp}$. This expression displays the numerator-based s term of a single-zero response.

Without bypass, a second response zero results from the inductances of the power-supply lines. Then the line inductances make the line impedances $Z_L = L_p s$, and substituting this expression in the last e_{noS} equation produces $e_{noS} = 2A_{ne}i_S L_p s^2/\omega_{cp}$. This expression displays the numerator-based s^2 factor of a double-zero response. At higher frequencies, such a response produces a very large gain for the power-supply noise signal. In the unbypassed case, reducing this gain requires decreasing L_p or increasing ω_{cp}, or both. Decreasing L_p reduces the supply-line noise signal, and increasing ω_{cp} improves the PSRR ability to reject this noise. These noise reduction options combine with power-supply bypass in the control of power-supply noise coupling.

Adding the C_B bypass capacitors of Fig. 3.1 dramatically reduces this noise coupling by removing both zeros of the e_{noS} response. Ideally, capacitive bypass dominates the higher-frequency line impedances, converting them from $Z_L = L_p s$ to $Z_L = 1/C_B s$. This removes a zero and adds a pole. Making this $Z_L = 1/C_B s$ substitution in the $e_{noS} = 2A_{ne} i_S Z_L s/\omega_{cp}$ equation produces $e_{noS} = 2A_{ne} i_S / C_B \omega_{cp}$. Here no s term appears in either the numerator or the denominator, reflecting the flat response normally expected for op amp noise. Then the power-supply noise produced by the signal current i_S, and coupled to the amplifier input through PSRR, only receives the same amplification as the amplifier's input noise voltage. However, this bypass must also ensure frequency stability, and bypass deficiencies still require attention to the noise reduction options involving reduced L_p and increased ω_{cp}, as described.

3.1.3 Power-supply coupling and frequency stability

In addition to noise reduction, the power-supply bypass must serve a second and generally more prevailing purpose, the preservation of frequency stability. A lack of stability or oscillation produces the ultimate noise signal, independently sustaining itself and easily dominating the circuit's response error. As a result, the bypass selection generally focuses first upon stability and then upon improvements that further reduce noise coupling. Stability preservation requires bypass control of the parasitic feedback loop established by the supply-line impedance and the amplifier's PSRR coupling. Analysis of this feedback loop provides insight into the loop's origin and control, as well as producing a mathematical expression of the condition required for oscillation.

As described with Fig. 3.1, this loop originates in the power-supply noise signal $2i_S Z_L$ that couples to the amplifier input in the error signal $e_{idS} = 2i_S Z_L / \text{PSRR}$. From there the amplifier produces an output signal and a corresponding feedback error signal that potentially supports e_{idS}. Figure 3.3 illustrates this feedback relationship with a modified version of the circuit in Fig. 3.1. As before, the circuit here neglects the input signal to focus the analysis upon the PSRR error signal and its feedback relationship. Modifications here also alter the power-supply connection and the input error signal. For the supply connection, the amplifier load current i_L replaces the previous i_S, and the V_+ connection alone illustrates the supply coupling effect. Only the amplifier's i_L portion of i_S influences stability because only this portion produces a feedback relationship with the amplifier. Also, the coupling effects described here for the V_+ connection apply to those

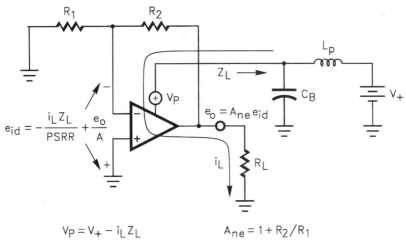

Figure 3.3 Load current i_L, drawn by the op amp, reacts with supply-line inductance to produce input error signal e_o/A, which in turn produces an output voltage and a potential for oscillation.

that occur with the V_- path as well. Next, a change in the input error signal shown reflects the effect of the amplifier gain error. In the figure, the e_o/A gain error term appears in the more general input error signal e_{id} instead of the previous e_{idS}, to include the parasitic feedback mechanism.

3.1.4 Oscillation condition

Tracing a signal through the parasitic feedback loop illustrates both the feedback mechanism and its potential for oscillation. An intuitive evaluation illustrates the mechanism and then a quantitative analysis defines the oscillation condition. In the figure, an e_o output voltage from the amplifier supplies a load current i_L to resistor R_L. The amplifier draws this current from V_+ and through the Z_L impedance of its supply line. The resulting line voltage drop produces a component of the e_{id} error signal $-i_L Z_L/\text{PSRR}$ through the amplifier's finite PSRR. Then the circuit amplifies this component by the circuit's noise gain A_{ne}, producing an e_o output response. That response reflects back to the amplifier inputs through the amplifier's open-loop gain A, producing the e_o/A component of the e_{id} shown. Thus the power-supply coupling produces an input signal which in turn produces an output signal that then produces an input signal. This simple link describes the full circle of a feedback loop capable of sustaining oscillation.

Quantifying this analysis produces two equations that describe the amplifier's feedback response and express the fundamental condition required for oscillation. First the direct and secondary effects of power-supply coupling produce the input error signal $e_{id} = -i_L Z_L/\text{PSRR} + e_o/A$, as described. Substituting $i_L = e_o/R_L$ removes the i_L variable for

$$e_{id} = \left(-\frac{Z_L}{R_L \cdot \text{PSRR}} + \frac{1}{A}\right)e_o$$

To sustain oscillation, e_{id} must independently support e_o through a second relationship. An op amp circuit amplifies its input error as expressed by[4]

$$e_o = \frac{A}{1 + A\beta} e_{id} = A_{ne} e_{id}$$

where $A_{ne} = A/(1 + A\beta)$ is the circuit's noise gain. The last two equations describe first the e_{id} dependence upon e_o and then a converse relationship. Oscillation results when the two signals support each other with no additional assistance. Combining the equations for the two produces the oscillation-defining condition

$$e_o = e_o \frac{Z_L A - R_L \cdot \text{PSRR}}{(1 + A\beta)R_L \cdot \text{PSRR}}$$

While too complex for intuitive evaluation, this last equation quickly yields two simplified results. With e_o on both sides of the equation, only two solutions satisfy the equality expressed. First, $e_o = 0$ balances the equation, indicating the stable state with no self-sustaining oscillation signal. Second, dividing both sides of the equation by e_o and solving for Z_L defines the line impedance required to produce the oscillation state,

$$Z_L = \frac{R_L \cdot \text{PSRR}}{A/(2 + A\beta)} \approx \frac{R_L \cdot \text{PSRR}}{A_{ne}}$$

Preventing oscillation requires avoiding conditions that balance this new equality. Keeping Z_L low avoids oscillation as long as

$$Z_L < \frac{R_L \cdot \text{PSRR}}{A_{ne}}$$

For the generalized example of Fig. 3.3, $A_{ne} = (1 + R_2/R_1)/(1 + f/f_c)$.

This last Z_L equation offers several insights into the bypass requirement and the related parameters that control stability. Variations in three parameters of the equation influence the Z_L requirement. At lower frequencies, bypass capacitors fail to reduce Z_L, increasing the value of the left side of the equation. However, at these low frequencies, PSRR remains high, making the right side of the equation large and preserving the equation's < condition. At higher frequencies, PSRR declines, decreasing the value of the right side. However, bypass capacitors simultaneously reduce Z_L to retain the stability condition. At all frequencies, the value of the load resistance R_L influences stability. As expressed before, lower values of R_L require correspondingly lower values of Z_L. From a circuit perspective, lower R_L values produce greater supply current drains, which increase the supply voltage drop developed on the Z_L impedance.

The foregoing analysis provides intuitive but simplified guidelines to stability preservation through control of the line impedance's magnitude. In practice, phase shifts also influence stability, and complex phase conditions actually accompany the Z_L magnitude requirement. The PSRR roll off adds a 90° phase shift and the bypassed Z_L impedance displays multiple resonances that potentially produce 180° phase transitions. These transitions compromise stability in a manner not evident from cursory examination of the Z_L equation. The complexity of the overall result presents a formidable analysis task, again best avoided by design practices intuitively communicated by that Z_L equation. Then examination of the Z_L resonances produces simple design equations that ensure stability.

3.2 Selecting the Primary Bypass Capacitor

Power-supply bypass does not simply replace the $Z_L = L_p s$ impedance with $Z_L = 1/C_B s$, as previously approximated. This approximation fails due to a variety of resonances formed with parasitic inductances. Two L-C resonances result from just the basic power-supply bypass condition. There the line inductance and the bypass capacitor produce the first resonance and the capacitor itself produces the second.

3.2.1 Fundamental bypass resonance

The first bypass resonance results from the reaction of C_B with L_p and occurs well within the frequency range of the amplifier. In Fig. 3.3 (p. 79), C_B effectively appears in parallel with the L_p supply-line inductance. There the ideal zero impedance of V_+ provides a direct return to ac ground, completing the parallel connection. Then C_B and L_p form the classic L-C tank circuit with an impedance of

$$Z_L = \frac{L_p s}{1 + L_p C_B s^2}$$

At low frequencies $Z_L \cong L_p s$ and at high frequencies $Z_L \cong 1/C_B s$, as previously approximated. However, at an intermediate frequency, this impedance displays a resonance maximum at the frequency $f_{rL} = 1/2\pi\sqrt{L_p C_B}$. There the Z_L line impedance approaches infinity, totally counter to the impedance reduction intended by the power-supply bypass.

Fortunately, a parasitic line resistance R_p and the amplifier's PSRR contend with this resonance. The R_p resistance dissipates the resonant energy of the L-C tank, detuning it to dramatically restrain the resulting impedance rise. A typical printed-circuit trace introduces a parasitic 12 mΩ per inch in its connection between the power supply and the amplifier. In addition, choosing C_B to place this resonance at a lower frequency where PSRR remains high lets the amplifier attenuate the associated coupling effect. Figure 3.4 models the detuned L-C tank circuit of the basic bypass condition. There bypass capacitor C_B shunts the series combination of the parasitic line inductance L_p and the parasitic line resistance R_p. This model assumes the zero output impedance of an ideal power supply and returns the L_p, R_p combination to ground. Then the circuit model represents the Z_L line impedance seen from the V_P terminal of the amplifier's positive supply connection. That circuit potentially supports the recirculating current i_r shown in a damped but ringing response.

Initially neglecting R_p of the model in Fig 3.4 illustrates the resonant condition. Then the parallel-connected L_p and C_B support the same voltage drop. At the resonance, the equal impedance magnitudes of L_p and C_B develop currents of equal magnitudes from their common voltage drop. Yet the 180° phase difference for the two impedances makes the two currents out of phase or of opposite polarity. This produces the circulating, resonant current i_r shown with no need for additional energy input, that is, oscillation. Current i_r flows through C_B in one direction and through L_p in the other, yet the two flows develop the same voltage drop, as required by the parallel connection of these elements. Thus a self-sustaining i_r loop results, in which energy trans-

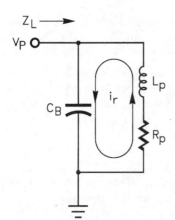

Figure 3.4 Together power-supply-line inductance and bypass capacitance form an L-C tank circuit with a resonant impedance detuned by the presence of parasitic line resistance R_p.

fers from the inductor to the capacitor and back, developing a sinusoidal voltage signal on Z_L at the resonant frequency f_r.

Fortunately, the presence of parasitic resistance R_p introduces energy dissipation to reduce the resonant impedance and stop the oscillation. Analysis of the figure's simple R-L-C combination shows that the line impedance peaks at the resonance frequency f_r. There Z_L displays the 0° phase shift of a resistive impedance and a resonant magnitude of

$$Z_{Lr} = \frac{\sqrt{L_p(R_p^2 C_B + L_p)}}{C_B R_p}$$

Examination of this equation confirms anticipated effects. Z_{Lr} requires nonzero values for both C_B and R_p to prevent an infinite peak impedance. Also, making C_B large decreases this impedance and reduces its resonant impedance equation to $Z_{Lr} = \sqrt{L_p/C_B}$.

3.2.2 Graphical evaluation of bypass resonance

Graphical evaluation of the Z_L line impedance versus frequency characterizes the resonant condition of the impedance in Fig. 3.5. There the frequency response of the Z_L impedance displays its characteristic resonant shape. Superimposed on the figure, the amplifier's A_{OL} response provides a relative frequency comparison of practical bypass conditions. As shown by the Z_L response, the line impedance varies from inductive to resistive to capacitive as the frequency increases. At lower frequencies, line inductance L_p controls Z_L, producing a rising impedance curve. At higher frequencies, bypass capacitor C_B takes control, producing the declining Z_L curve desired. Between these two frequency ranges, the inductive and capacitive response curves cross at their resonant frequency f_r. There Z_L rises to a peak value and actually increases due to the presence of bypass capacitor C_B. Fortunately, the presence of resistance R_p limits this rise by detuning the tank circuit.

Examination of the component curves that define the Z_L response in Fig. 3.5 explains the resonant condition in familiar response plot terms. As mentioned, inductive and capacitive influences largely shape the Z_L response with one or the other dominant at frequencies separated from the resonance at f_r. However, in the vicinity of f_r, the magnitude and phase relationships of the inductive and capacitive components combine to increase the Z_L line impedance. This effect of this combination parallels the magnitude-phase relationship of the basic op amp stability analysis described in Chap. 1. In both cases, the crossing of two magnitude response curves marks the frequency of magnitude equality since the two curves then occupy the same point

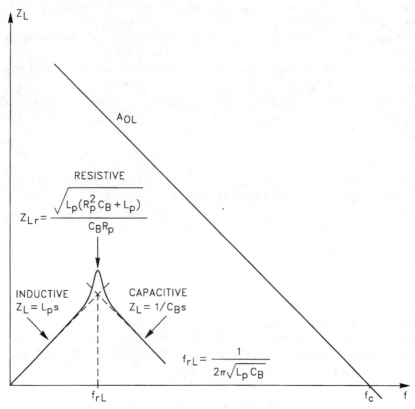

Figure 3.5 As frequency increases, bypassed line impedance varies from inductive to capacitive, developing a resonant peak at the transition point.

on a graph. In the figure, dashed-line extensions of the inductive and capacitive impedance responses mark this point, where the $Z_L = L_p s$ and $Z_L = 1/C_B s$ lines cross at f_r.

At this crossing, the slopes of the two lines reflect their associated phase shifts, and the slope difference, or rate of closure, reflects the corresponding phase difference. At f_r, this phase difference turns this equal-magnitude point into a resonance. The single-zero rising response of the $L_p s$ component reflects a 90° phase lead and the single-pole declining response of the $1/C_B s$ component reflects a 90° phase lag. This combination produces a 180° phase difference between the impedances of the L_p and C_B components of the circuit model. For the ideal L-C tank circuit, this magnitude-phase combination produces an infinite impedance, and any supply of energy to the tank produces oscillation. Simply turning on the power supply provides this energy through the transient current drawn from the supply. After that, the ideal tank's infinite impedance successfully sustains an oscillation signal, even after termination of the energy input to the circuit.

3.2.3 Bypass selection

Control of the resonance described in Sec. 3.2.2 and an empirical guideline produce the first two design equations for the bypass selection. The first equation places the resonance at a frequency where a higher amplifier PSRR better contends with the associated impedance rise. The second equation sets the magnitude of the bypassed line impedance with a C_B design equation.

At first the impedance increase of the Z_{Lr} resonant condition of Fig. 3.5 would suggest moving f_{rL} to a frequency above the normal operating frequency range f_c of the op amp. Here f_c represents the unity-gain crossover frequency of the op amp and generally represents the upper limit of the amplifier's useful frequency range. Above f_c the lack of amplifier gain restricts the parasitic feedback resulting from the resonance. However, making $f_{rL} > f_c$ would entirely sacrifice the benefit of the supply bypass. Preceding f_{rL} in Fig. 3.5, the Z_L impedance rises with frequency, unaffected by the bypass. The Z_L impedance only reflects the desired decline after the f_{rL} resonance where the capacitive roll off dominates Z_L. Thus the f_{rL} resonance must be endured within the op amp's operating frequency range in order to benefit from the higher-frequency effects of the bypass.

Moving this bypass resonance to a frequency $f_{rL} \ll f_c$ actually provides the best compromise. This places the resonance at a frequency where the amplifier retains sufficient PSRR to attenuate the coupling effect of the resonance. Further, the $f_{rL} \ll f_c$ placement moves the C_B roll off of the bypass back into the amplifier's useful frequency range. Then for $f_{rL} = 1/2\pi\sqrt{L_p C_B} \ll f_c$ the C_B capacitance value must satisfy the limit equation

$$C_B \gg \frac{1}{4L_p(\pi f_c)^2}$$

Generally, the line impedance control described next automatically satisfies this C_B requirement.

Making $f_{rL} \ll f_c$ also reduces the analysis of other bypass conditions to the impedance of the capacitor alone. Without bypass, the line impedance actually includes a complex combination of inductances and resistances introduced by wire, printed-circuit traces, and connectors. All of these affect the lower-frequency response of the Z_L impedance, discouraging any detailed analysis. However, making $f_{rL} \ll f_c$ reduces the Z_L impedance to just that of the capacitor in the critical higher-frequency range. There a declining PSRR most requires a low Z_L to limit supply-line coupling. Above the f_{rL} resonance, Z_L then becomes $Z_L = Z_{CB} = 1/C_B s$, independent of the preceding line impedance complexity.

Then for most op amps, simply reducing the C_B impedance to about 1 Ω well before f_c effectively counteracts the line impedance effects. This reduces the Sec. 3.1.4 stability condition to $Z_L = Z_{CB} = 1 < R_L \cdot$ PSRR/A_{ne} or $A_{ne} < R_L \cdot$ PSRR. Here a noise gain as large as 1000 and a load resistance as small as 1 kΩ permit a PSRR as low as unity before oscillation results. Rarely would such a low PSRR occur where the noise gain remained this high, so the $Z_{CB} = 1$ Ω guideline assures stability for virtually every case. Setting $Z_{CB} = 1/2\pi f C_B = 1$ Ω produces $C_B = 1/2\pi f$. Then selecting $f = f_c/100$ for this 1 Ω condition produces a general design equation for C_B,

$$C_B = 50/\pi f_c$$

However, retaining the guideline 1 Ω impedance up to and beyond f_c may require additional bypass attention, as described hereafter.

3.3 Selecting a Secondary Bypass Capacitor[5]

Capacitor self-resonance potentially compromises the bypass described, requiring the addition of secondary bypass capacitors. For general-purpose op amps the bypass selection generally places this self-resonance at a frequency beyond the amplifier's response range. Higher-frequency amplifiers encompass this frequency and may require the additional bypass. Still, even those capacitors exhibit self-resonances and require careful selection through a compromise that again produces simple design equations.

3.3.1 Bypass capacitor self-resonance

The inherent parasitic inductances and resistances of capacitors also disturb the bypass effectiveness. The inductances introduce new resonances and the resistances limit the line impedance reduction. A first new resonance results just from the capacitance and inductance of a single bypass capacitor in a self-resonant condition. At this resonance, the bypass impedance would drop to zero, except for the presence of the parasitic resistance. Above that frequency, the capacitor's parasitic inductance overrides the intended capacitance in the control of the bypass impedance.

Making $f_{rL} \ll f_c$ typically requires a fairly large capacitance value in the multiple-microfarad range. Capacitors this large introduce a new

Power-Supply Bypass 87

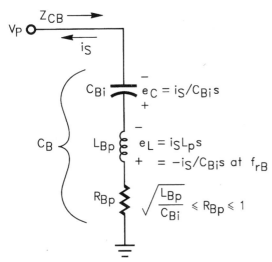

Figure 3.6 Practical bypass capacitors possess inductive and resistive parasitics that result in a detuned capacitor self-resonance.

resonance condition that compromises the bypass effectiveness at frequencies sometimes within the amplifier's response range. The new resonance results from the parasitic inductance L_{Bp} of the bypass capacitor reacting with the capacitor's ideal capacitance. Figure 3.6 models the bypassed line impedance, including the actual impedance elements of a practical capacitor. Here the circuit model simplifies to show only those elements pertinent to the bypass discussion. This model neglects the L_p and R_p line parasitics since the bypass capacitor controls the line impedance in the frequency range of interest here.

However, the model does include the capacitor's parasitic inductance and resistance. All capacitors possess these parasitics, with values corresponding largely to the lengths of the capacitor's external and internal connecting paths. Minimizing the total connecting length minimizes the parasitics and maximizes the frequency of the self-resonance. First, just minimizing the capacitor lead length, including that of associated circuit-board traces, addresses the external path length. Then capacitor selection minimizes the internal path component as primarily determined by the capacitor's construction material. For a given capacitance value, capacitor materials with higher capacitance densities reduce this length. Tantalum, NPO ceramic, and mica capacitors offer the lower parasitic alternatives for high, medium, and low capacitance values, respectively.

More detailed examination of the bypass capacitor's actual impedance components provides insight into its nonideal behavior and the means to control the bypass result. As modeled in Fig. 3.6, L_{Bp} produces a self-resonance with C_B's ideal capacitance C_{Bi}, and the parasitic resistance R_{Bp} sets the capacitor's impedance at the resonance. The supply current drain i_S develops voltages across each of the model's impedances, with the capacitive and inductive impedances producing canceling effects at their resonance. Current i_S produces the voltage drops $e_C = i_S/C_{Bi}s$ across C_{Bi} and $e_L = i_S L_{Bp} s$ across L_{Bp}. At the resonance, $s = s_{rB} = j\omega_{rB} = j/\sqrt{L_{Bp}C_{Bi}}$. Substituting this result for s in the preceding voltage drop equations yields $e_{Cr} = -j\sqrt{L_{Bp}/C_{Bi}}$ and $e_{Lr} = j\sqrt{L_{Bp}/C_{Bi}}$. Thus at resonance, $e_{Cr} = -e_{Lr}$ and the two voltage drops cancel. Except for the presence of R_{Bp}, this voltage cancellation would produce zero voltage drop across the capacitor in response to the i_S current, indicating the ideal zero-bypass impedance at the resonance.

However, parasitic resistance R_{Bp} produces a third voltage drop with i_S and limits the impedance decline to the level of this resistance. At first this limit would seem undesirable, but the parasitic resistance also detunes the C_{Bi}, L_{Bp} resonance, taming a dramatic phase transition in the line impedance. That transition presents a broad range of phase conditions that could degrade stability. In practice, this effect rarely produces a significant stability disturbance because it decreases rather than increases the line impedance. However, the impedance rise following the resonance may well degrade stability and require dual bypass, as described here. For the general case, the beneficial effect of the capacitor's parasitic resistance provides adequate stability assurance and eases the capacitor selection. Accepting a capacitor with higher parasitic resistance actually smoothes the phase transition of the resonance.

This resistance benefits performance as long as it detunes the resonance but does not raise the line impedance above the 1-Ω guideline level. For optimum detuning, the impedances of the R, L, and C components of the tank circuit should be equal at the resonant frequency. At this frequency, the impedance magnitudes of C_{Bi} and L_{Bp} automatically equate, as shown in Fig. 3.7. There dashed-line extensions of the two impedance curves identify equal magnitudes at their f_{rB} crossing. At this crossing, $Z_{CBi} = 1/2\pi f_{rB} C_{Bi}$ where $f_{rB} = 1/2\pi\sqrt{L_{Bp}C_{Bi}}$. Setting R_{Bp} equal to the Z_{CBi} impedance and solving produces the minimum resistance required for detuning, $R_{Bp} = \sqrt{L_{Bp}/C_{Bi}}$. Combining this minimum with the 1 Ω maximum produces the design limit equation

$$\sqrt{\frac{L_{BP}}{C_{Bi}}} \leq R_{Bp} \leq 1$$

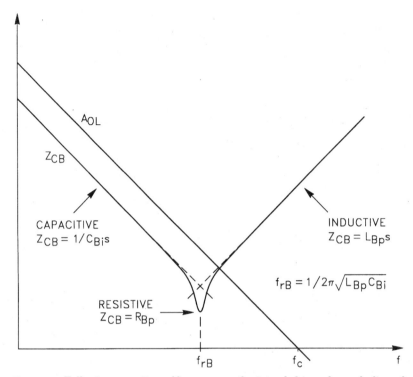

Figure 3.7 Following capacitor self-resonance, the intended impedance decline of a bypass capacitor reverses when capacitor internal inductance takes control.

Op amps having lesser signal bandwidths generally assure this detuned condition in the bypass capacitor selection. Such amplifiers require larger C_B values for the previous C_B design equation, directing the capacitor selection to tantalum types. Such capacitors display larger R_{Bp} values that flatten the Fig. 3.7 transition between its capacitive and inductive slopes. However, for higher-frequency amplifiers the C_B design equation would permit the use of ceramic capacitors which display much lower R_{Bp} values. Then in the event of an unexplained stability degradation, replacing an initial ceramic choice with tantalum could provide the solution.

Following the f_{rB} resonance, the Z_{CB} impedance curve rises in Fig. 3.7, again threatening stability. However, the C_B selection guidelines described control the effect of this rise for general-purpose op amps adequately. There the avoidance of excessively large capacitance values and the limited amplifier bandwidths reduce the susceptibility to the impedance increase. Choosing $C_B = 50/\pi f_c$ assures adequate line impedance control at lower frequencies without introducing the excess inductance of a larger capacitor. Success here also depends upon minimizing the connection length between the capacitor and the amplifier. Then the Z_{CB} impedance rise remains within the 1 Ω limit over the useful bandwidth of general-purpose amplifiers. However, op amps having greater bandwidths generally require the addition of secondary bypass capacitors, as described hereafter.

3.3.2 Dual-bypass capacitors

The simple single-capacitor bypass that serves general-purpose op amps fails with higher-frequency op amps. At higher frequencies, inductive effects become far more serious, requiring greater attention to frequency stability and supply noise coupling. The parasitic inductance of the larger-value single-bypass capacitor defeats the bypass purpose at higher frequencies, requiring the addition of a smaller secondary capacitor. Even general-purpose amplifiers may require dual bypass if PSRR is low.

Higher-frequency op amps expand the meaningful extent of the $Z_L = L_{Bp}s$ impedance rise in Fig. 3.7, jeopardizing stability and increasing supply noise coupling. There greater amplifier bandwidths encompass more of the high-frequency end of this rise and require secondary bypass to counter the inductance of the primary bypass. Adding a smaller capacitor in parallel with the first commonly bypasses this new inductance limit. However, simply adding the secondary bypass does not automatically resolve the line impedance issue. The secondary capacitor has an inductance of its own, producing another bypass compromise at a somewhat higher frequency. Further, the secondary capacitor reacts with the inductance of the first, producing a resonant increase in the supply-line impedance. Careful selection of this capacitor produces a compromise solution that retains low bypass impedance throughout the amplifier's significant frequency range. For this selection, analysis of the net bypass impedance produces two simple design equations that guide the capacitor selection.

Adding a secondary bypass capacitor in parallel with the first restores a low bypass impedance for the full response range of most high-frequency amplifiers. This produces a net $C_B = C_{B1} + C_{B2}$, and making $C_{B2} \ll C_{B1}$ generally assures that the C_{B2} self-resonance occurs outside the amplifier's response range. Both the lower capacitance value of C_{B2} and the accompanying lower parasitic inductance produce this higher resonant frequency. For even higher-frequency applications yet a third capacitor may be required. In either case, examination of the dual-bypass case here illustrates the impedance characteristics that guide the bypass selection process.

Adding secondary bypass capacitors restores the declining frequen-

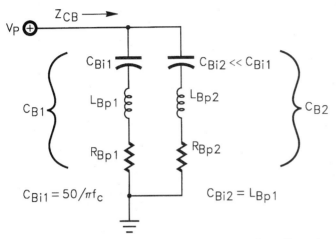

Figure 3.8 Adding a secondary bypass capacitor of smaller value bypasses inductance of the first but also produces a resonance with that inductance and introduces a new self-resonance.

cy response of the Z_{CB} bypass impedance, but with new complications. This solution introduces two additional resonances, one from the self-resonance of the secondary capacitor and the other from the reaction of the secondary capacitance with the inductance of the first capacitor. Figure 3.8 models the supply-line impedance for the dual-bypass circuit with the same simplifications as those used for the single-bypass case. As before, this model neglects the L_p and R_p line parasitics since only the capacitors affect the Z_L response in the frequency range of interest here. Thus the model represents the supply line by just the ideal capacitances and their associated parasitics. These parasitics introduce two new resonances for the dual-bypass case. The first results from the self-resonance of C_{B2} and the second from the reaction of C_{B2} with L_{Bp1}.

Figure 3.9 illustrates the first of these along with the benefit of the dual bypass. Two curves in Fig. 3.9 display the resulting capacitor impedances as Z_{CB1} and Z_{CB2}. At lower frequencies, a declining Z_{CB1} provides the lower-impedance bypass shunt. However, Z_{CB1} later resonates and begins to rise at f_{rB1}. At higher frequencies, Z_{CB2} bypasses this rise and initially restores a declining bypass impedance. Later, the self-resonance of C_{B2} at f_{rB2} again produces a rise, but at a lower impedance level than provided by the Z_{CB1} curve. While the resonances remain inevitable, their careful placement optimizes the overall bypass effect. To aid this placement, a third curve of the graph shows the amplifier's open-loop gain response A_{OL}. Comparison of the graph's three responses guides the secondary capacitor selection in a manner that ensures the amplifier's frequency stability.

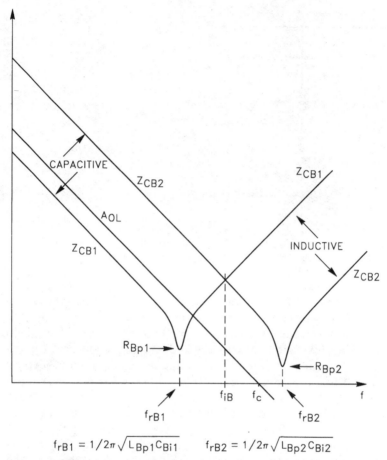

Figure 3.9 With dual bypass, the added C_{B2} capacitor restores low line impedance following the C_{B1} resonance, but only up to the C_{B2} self-resonance.

$$f_{rB1} = 1/2\pi\sqrt{L_{Bp1}C_{Bi1}} \qquad f_{rB2} = 1/2\pi\sqrt{L_{Bp2}C_{Bi2}}$$

3.3.3 Dual-bypass selection

Selection of the primary capacitor C_{B1} indirectly defines the C_{B2} capacitance value through impedance interaction. The fundamental bypass requirement defines C_{B1} as described before for the single-bypass case; $C_{B1} = 50/\pi f_c$. This selection reduces the line impedance to 1 Ω at a frequency well within the amplifier's response range. Then the addition of C_{B2} bypasses the higher-frequency impedance rise produced by the L_{Bp1} inductance of C_{B1}. Selecting C_{B2} to limit this rise to the same 1-Ω guideline level defines this secondary capacitor's value.

In Fig. 3.9 the net bypass impedance of the C_{B1}, C_{B2} parallel combination reaches one of its maxima at the f_{iB} intercept of the rising Z_{CB1} curve and the falling Z_{CB2}. For higher-frequency amplifiers this peak would typically occur within the amplifier's response range. Selecting a value for C_{B2} that limits this maximum to 1 Ω continues the stability preservation described before for the C_{B1} selection. At their f_{iB} intercept, the two Z_{CB} curves occupy the same point on the graph, making $Z_{CB2} = Z_{CB1}$. There inductance controls $Z_{CB1} = 2\pi f_{iB} L_{Bp1}$ and capacitance controls $Z_{CB2} = 1/2\pi f_{iB} C_{Bi2}$. Equating the two impedance expressions defines the intercept frequency as

$$f_{iB} = 1/2\pi\sqrt{L_{Bp1} C_{Bi2}}$$

Thus the parasitic inductance of C_{B1} and the ideal capacitance of C_{B2} combine to define the frequency of this bypass impedance maximum.

Translating this f_{iB} expression into a C_{B2} design equation requires one further step. At f_{iB}, the Z_{CB2} curve follows its capacitive roll off as expressed by $Z_{CB2} = 1/2\pi f C_{B2}$. Setting $Z_{CB2} = 1$ Ω and $f = f_{iB}$ in this expression produces $f_{iB} = 1/2\pi C_{B2}$. Equating this result with the f_{iB} expression yields the C_{B2} design equation

$$C_{B2} = L_{Bp1}$$

Thus simply making the magnitude of the C_{B2} capacitance equal to that of C_{B1}'s parasitic inductance transfers line impedance control from Z_{CB1} to Z_{CB2} at the 1-Ω level. In practice a resonance described later raises the peak impedance above 1 Ω at f_{iB}, but temporarily ignoring this effect yields the best starting point for the bypass selection. As described later, resistive detuning of this resonance or an alternate capacitor type solves any related problem without altering the design equation.

The C_{B2} equation requires knowing L_{Bp1}, and impedance measurement provides the most reliable determination of this inductance. Capacitor data sheets rarely specify this parasitic, although some data sheets indirectly communicate this inductance through an impedance-versus-frequency curve. There the higher-frequency rising portion of this curve yields the parasitic inductance through the simple calculation $L_p = Z_C/2\pi f$. Unfortunately, even these curves can fail to communicate the actual inductance condition of practical applications. For packaged capacitors, this resonance includes the parasitic inductance of package leads that are largely removed upon installation. Chip capacitors avoid this lead effect, but their response curves still do not include the effect of the circuit-board traces that connect the capacitor. Fortunately, today's impedance analyzers measure the frequency response of a given capacitor accurately under the conditions of its approximated connection length. From such measurements, the rising portion of the impedance curve then accurately defines the actual inductance through the $L_p = Z_C/2\pi f$ relationship.

3.4 Bypass Alternatives

The preceding discussion outlines the bypass selection that serves most applications adequately. However, three special conditions may require additional attention to supply-line coupling. For these conditions, a third bypass capacitor, filtering, or resistive detuning further reduces the coupling effects. The third capacitor alternative extends the frequency range of bypass control for even higher-frequency amplifiers. There the amplifier's A_{OL} response can encompass the self-resonances of both C_{B1} and C_{B2}. Then adding a third, and even smaller, bypass capacitor rolls off the inductive impedance of C_{B2}, just as C_{B2} did for C_{B1}. Amplifiers with bandwidths exceeding 30 MHz may require this third capacitor.

A second alternative, filtering, removes supply-coupled signals outside the useful PSRR range of the amplifier. There coresident circuitry may introduce high-frequency supply-line signals that couple through the amplifier with little PSRR attenuation. Beyond the PSRR's unity-gain crossover, such signals typically couple straight through the amplifier to its output. However, this PSRR crossover typically resides near the amplifier's A_{OL} crossover, presenting a filtering opportunity. The A_{OL} crossover marks the end point of the amplifier's useful frequency range. Thus low-pass filtering, applied after the amplifier, can often remove the effect of high-frequency PSRR coupling without restricting the useful bandwidth of the application. Still, intermodulation effects can circumvent the filtering, as described later.

3.4.1 Detuning the dual-bypass resonance

Resistive detuning of the bypass impedance offers a third alternative for the reduction of supply-coupling effects. The dual-bypass configuration potentially produces a critical resonance that degrades stability at surprising frequencies. This may occur at a frequency below or above the amplifier's response crossover at f_c. Below f_c the C_{B2} selected restores a line impedance below the 1-Ω guideline level, presumably preserving stable operation. However, the C_{B2} resonance with L_{Bp1} can raise the net line impedance well above this level, producing oscillation. In other cases, the resonance can even produce oscillation at frequencies above f_c. There diminished amplifier gain limits the parasitic feedback loop and would seem to prevent instability. However, the resonant impedance rise can counteract this limit.

When necessary, resistive degeneration detunes this new resonance and an appropriate capacitor selection generally provides this just through the capacitor's parasitic resistance.

An interaction between the two capacitors of the dual-bypass configuration develops this new resonance. There the secondary capacitance reacts with the inductance of the primary in another L-C tank configuration. Figure 3.10 illustrates this effect with the bold Z_{CB} response curve representing the net impedance of the parallel-connected bypass capacitors. There Z_{CB} first follows the Z_{CB1} curve of C_{B1} at lower frequencies and later follows the Z_{CB2} curve of C_{B2}. The transition between the two occurs at their f_{iB} intercept where the bypass combination produces a resonance of its own. This resonance increases the Z_{CB} magnitude as shown and produces a 180° transition in the

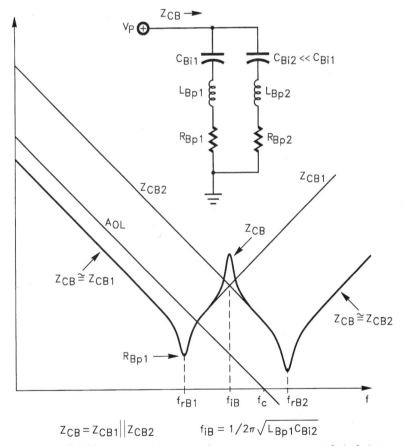

Figure 3.10 Dual-bypass capacitors introduce a new resonance at their f_{iB} intercept, where C_{Bi2} reacts with L_{Bp1}, increasing net Z_{CB} impedance.

phase of Z_{CB}. The magnitudes and phases of the amplifier's PSRR and A_{OL} responses combine with these Z_{CB} characteristics to produce a complex parasitic feedback condition.

Some combinations of application conditions make the magnitude or the phase of this resonance, or both, degrade stability. In such cases, attention to the more conventional stability determinant, the amplifier's 1/ß intercept, fails to identify the problem. Similarly, adding a third bypass capacitor produces no improvement. The third capacitor only reduces the line impedance at frequencies above this interaction resonance. For such cases, bypass detuning moderates the PSRR error signal by reducing both the magnitude and the phase shift of the supply-line impedance. Careful selection of the detuning resistance improves stability without increasing the supply-line impedance in the critical higher-frequency ranges.

Graphical evaluation of this f_{iB} resonance reveals its underlying causes and displays the cure provided by the resistive degeneration. The resonance results from the reaction of C_{B1}'s parasitic inductance with C_{B2}'s capacitance. At f_{iB} in Fig. 3.10, Z_{CB1} follows a rising impedance response as controlled by L_{Bp1}, and Z_{CB2} follows a declining response as controlled by C_{Bi2}. At the crossing of the two responses, resonance results due to the equal magnitudes and opposite phases of the Z_{CB1} and Z_{CB2} impedances. This crossing or intercept of the two curves marks the point of equal impedance magnitudes since both curves occupy the same point in the graph. There neither impedance dominates the parallel combination, in contrast to the impedance conditions at points separated from f_{iB}. At this intercept, the equal magnitudes of Z_{CB1} and Z_{CB2} contribute equally to the combined Z_{CB} impedance, making the phase shifts of the two also equally significant. The single-zero rise of the Z_{CB1} curve corresponds to a 90° phase lead, and the single-pole decline of Z_{CB2} corresponds to a 90° phase lag. The 180° phase difference and equal impedance magnitudes inherently produce a resonance.

In Fig. 3.11, adding a small resistance R_{D1}, in series with C_{B1} detunes this resonance to ensure stability. In practice, just choosing a different capacitor type for C_{B1} often serves this purpose through the capacitor's parasitic resistance. This added resistance does increase the Z_{CB} magnitude in a lower-frequency range, but it then decreases Z_{CB} in the higher, more critical frequency range of f_{iB}. As will be seen, R_{D1} also serves to detune the self-resonance of C_{B1}.

Figure 3.12 illustrates the bypass optimum achieved with the degeneration resistance R_{D1}. As in Fig. 3.10, the bold Z_{CB} curve begins by following Z_{CB1} and ends by following Z_{CB2}. Unlike in Fig. 3.10, the Z_{CB} curve now makes a gradual rather than a resonant transition between Z_{CB1} and Z_{CB2} at their f_{iB} intercept. The addition of R_{D1} actu-

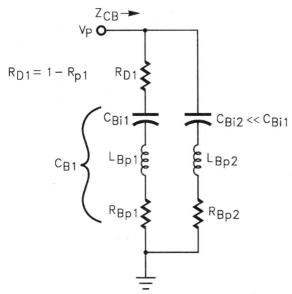

Figure 3.11 Adding a small resistance R_{D1}, in series with C_{B1} detunes interaction resonance between two capacitors.

ally detunes two resonances—first the self-resonance of C_{B1} and then the interactive resonance of C_{B1} and C_{B2}. Comparison of Fig. 3.12 with Fig. 3.10 reveals the overall effect of this resistance. Previously, the self-resonance of C_{B1}, at f_{rB1} in Fig. 3.10, produced a resonant impedance drop to the level of the parasitic resistance R_{Bp1}. In Fig. 3.12, the addition of R_{D1} removes this drop and raises this level to $R_{D1} + R_{Bp1}$. This actually raises the bypass impedance in the region of the previous f_{rB1} resonance. However, the amplifier generally retains a high PSRR in this frequency range for sufficient attenuation of supply-line effects. Further, the Z_{CB} curve now makes a smooth rather than a resonant transition to this new limit level. There the reduced Z_{CB} response slope indicates a greatly reduced phase transition at the frequency of the previous f_{rB1} resonance. This reduces the potential phase combinations with amplifier gain and PSRR effects that could occasionally degrade stability. Also, accepting the lower-frequency Z_{CB} increase allows R_{D1} to detune the more critical resonance at f_{iB}. There the Z_{CB} curve no longer displays a resonant peak, indicating a greatly reduced Z_{CB} magnitude and phase transition at f_{iB}.

3.4.2 Selecting the detuning resistance

The detuning resistance R_{D1} should be large enough to prevent the resonant impedance rise at f_{iB} but not so large as to unnecessarily

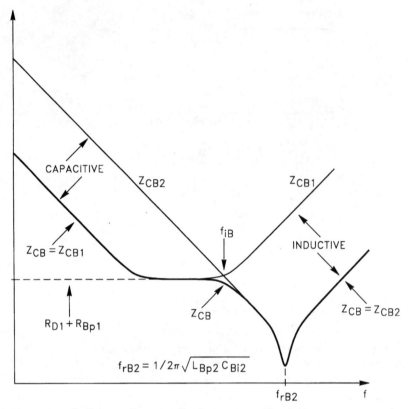

Figure 3.12 Resistance R_{D1} actually detunes two Fig. 3.10 resonances at the expense of an elevated Z_{CB} level at intermediate frequencies.

raise the Z_{CB} impedance at lower frequencies. Graphical analysis of Fig. 3.12 reveals this optimum condition and defines an R_{D1} design equation. The curves of the figure identify this optimum through the magnitudes and slopes of the impedance responses.

Detuning the Fig. 3.12 resonance at f_{iB} requires reducing the slope difference, or rate of closure, of the Z_{CB1} and Z_{CB2} curves at their f_{iB} intercept. This rate-of-closure criterion follows from the Chap. 1 stability analysis, where the slopes of intersecting response curves reflect phase differences and the potential for circuit oscillation. Here the slopes of the Z_{CB1} and Z_{CB2} curves similarly reflect phase differences and the potential for resonance. Adding R_{D1} decreases this slope difference at the f_{iB} intercept of Fig. 3.12, detuning the resonance. This added resistor produces the level zero-slope portion of the Z_{CB1} response, and this slope reflects zero phase shift. At the f_{iB} intercept, Z_{CB2} still follows its single-pole roll off, reflecting a 90° phase lag for a net 90° phase difference at the intercept. This greatly reduces the phase difference from the 180° required for resonance and would absolutely remove the resonant reaction of Z_{CB1} and Z_{CB2}.

However, bypass optimization produces a compromise favoring a somewhat greater phase difference at f_{iB}. The $R_{D1} + R_{Bp1}$ resistance, which produces the level region of the Z_{CB1} response, eventually looses control, returning the response curve to a single-zero rise. In the example in Fig. 3.12, the L_{Bp1} impedance overrides that of R_{D1} just at the f_{iB} intercept. Phase shift accompanies the following rise, and intuition first suggests increasing the R_{D1} resistance to move the Z_{CB1} rise, as well as its phase shift, well beyond the f_{iB} intercept. However, such a choice would increase the impedance magnitude of Z_{CB}'s new level region at $R_{D1} + R_{Bp1}$ unnecessarily. There a continued increase of R_{D1} would eventually increase rather than decrease the impedance at f_{iB}. In compromise, setting the $R_{D1} + R_{Bp1}$ level at the intercept magnitude still detunes the resonance while restraining the impedance increase of the level region.

As illustrated, this optimum level places the zero of the Z_{CB1} rise at the f_{iB} intercept. Then Z_{CB1} introduces 45° rather than the ideal 0° of phase shift at the intercept. This increases the net phase difference with Z_{CB2} from 90° to 135° but still avoids the 180° that produced the resonance before. For this compromise, $R_{D1} + R_{Bp1} = Z_{CB2}$ at f_{iB}. Then $R_{D1} + R_{Bp1} = 1/\omega_{iB}C_2$, where $\omega_{iB} = 2\pi f_{iB} = 1/\sqrt{L_{Bp1}C_{B2}}$. Combining these expressions and solving for R_{D1} yields $R_{D1} = \sqrt{L_{Bp1}/C_{B2}} - R_{Bp1}$. From before, the 1-Ω impedance guideline defines $C_{B2} = L_{Bp1}$, and substitution yields the design equation

$$R_{D1} = 1 - R_{Bp1}$$

Expressed differently, this equation prescribes $R_{D1} + R_{Bp1} = 1$ Ω for a total 1-Ω resistance in the C_{B1} path. This repeats the 1-Ω guideline underlying bypass selection.

Typically, this R_{D1} equation prescribes difficult resistance values of a fraction of an ohm. Such resistors exist but can be expensive and more difficult to acquire. However, when required, only this solution removes the effects of the f_{iB} bypass resonance. Fortunately, just choosing a different capacitor for C_{B1} generally provides an adequate detuning resistance. Choosing a C_{B1}' with $R_{Bp1}' \approx 1$ Ω detunes the f_{iB} resonance but still avoids exceeding the 1-Ω guideline. Fortunately, this resistance does not have to be precise to provide sufficient detuning of the f_{iB} resonance. Often just switching C_{B1} from a ceramic type to tantalum serves this purpose.

Note that this C_{B1} switch also requires changing the secondary bypass value, C_{B2}. A tantalum C_{B1}' substitutes an increased L_{Bp1}' for L_{Bp1}, requiring re-computation of the corresponding $C_{B2}' = L_{Bp1}'$.

Increasing C_{B2} here theoretically increases its inductance and would degrade higher-frequency stability. However, practical limitations generally leave the net inductance condition unchanged. In practice, some minimum lead length and inductance always accompany a capacitor construction and installation. Just the lead length from an amplifier package pin to the internal chip will be about 0.05 inch and introduces about 0.8 nH. Chip capacitors serving the secondary bypass function typically display similar inductances, independent of the capacitor value. Thus increasing C_{B2} to accommodate the C_{B1} change does not materially degrade stability.

3.5 Power-Supply Decoupling

A power-supply bypass provides the best first defense against supply-coupled noise and related instability. However, sometimes the bypass that preserves stability does not sufficiently reduce the noise coupled from other circuitry. Then supply decoupling filters may be required. This typically occurs when the supply-line noise contains frequencies above the amplifier's useful PSRR range. There the lack of PSRR often couples supply-line noise straight through the amplifier to its output. Subsequent filtering would remove this noise, except for the intermodulation effect of this higher-frequency noise. As described before, the filtering effectively removes the direct effect of this higher-frequency error signal without restricting the signal bandwidth. However, this filtering cannot similarly remove noise components that intermodulation distortion downshifts into the amplifier's useful frequency range. There the higher-frequency supply noise reacts with the amplifier's signal and distortion sources, producing error signals in the lower-frequency range of the amplifier.

3.5.1 Decoupling alternatives

Prevention supersedes cure in the reduction of supply-coupled noise. Power-supply decoupling filters prevent the higher-frequency signals from even reaching the amplifier. The simplest decoupling alternatives include R-C, L-C, or R-L-C low-pass filters, as illustrated in Fig. 3.13. There three low-pass filter configurations represent decoupling circuits placed between a power-supply source V_+ and the amplifier's positive supply pin V_P. While shown as a single capacitor, the C_B there represents the combined capacitance of all capacitors used for the supply bypass of V_P. In practice, the negative supply line would have an identical filter, but the positive supply case shown here serves to illustrate the decoupling principles.

Each of the three filter alternatives adds series impedance in the

Power-Supply Bypass 101

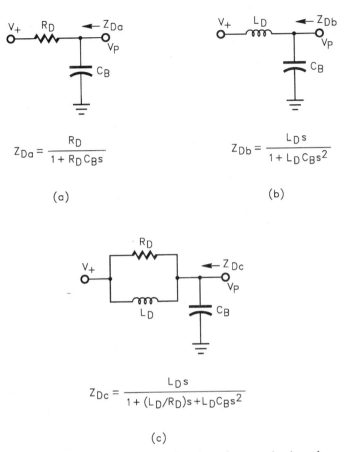

Figure 3.13 Power-supply decoupling introduces series impedance between the output of a power supply V_+ and op amp supply terminal V_P to intercept high-frequency supply-line transients.

supply line to combine with the bypass capacitance C_B, forming a low-pass filter. Then the high-frequency noise contained in V_+ drops across the series impedance rather than transferring to the V_P supply pin of the op amp. The series impedance added does not have to be mounted close to the amplifier to provide the desired decoupling. However, this impedance should be placed so that it does not conduct the supply current drain of circuitry other than the amplifier.

The type of impedance added determines the relative cost and decoupling effectiveness. This effectiveness depends upon both the filter response and the line impedance presented to the amplifier. The R-C solution of Fig. 3.13a offers the lowest cost and still produces appreciable decoupling results. With this circuit, the decoupling resistance R_D absorbs and dissipates the higher-frequency transient energy, transferring a more stable voltage to the C_B capacitance. Analysis of the decoupling circuit expresses this action in the classic R-C filter response

$$V_P = \frac{V_+}{1 + R_D C_B s}$$

This response rolls off the supply-line noise with a single pole at $1/2\pi R_D C_B$.

However, at lower frequencies this simple R-C decoupling increases the supply-line impedance, potentially increasing the PSRR error. To define this impedance effect, assume that the V_+ power supply presents zero output impedance, placing R_D effectively in parallel with C_B. This makes the decoupling impedance seen from the V_P terminal

$$Z_{Da} = \frac{R_D}{1 + R_D C_B s}$$

Thus at low frequencies the R-C filter increases the line impedance by the amount R_D. To limit this effect, the R_D resistance must be kept low, and a later discussion describes its value selection.

Alternately, L-C filtering removes this impedance increase and produces a double- rather than single-pole roll off of supply-line noise. However, this alternative introduces another impedance increase through a supply-line resonance. Figure 3.13b illustrates this case with inductor L_D simply replacing the previous resistor R_D. Then the L-C filter presents the low impedance of the inductor to lower-frequency current demands from the V_+ power supply. The filtering response of this alternative rolls off supply-line noise with a double pole at $f = 1/2\pi\sqrt{L_D C_B}$, as described by

$$V_P = \frac{V_+}{1 + L_D C_B s^2}$$

Compared with the R-C filter, the L-C filter reduces the supply-line impedance, but only at lower frequencies. The impedance seen from the V_P terminal becomes

$$Z_{Db} = \frac{L_D s}{1 + L_D C_B s^2}$$

At lower frequencies, the s term of the Z_{Db} numerator assures low impedance, but this term tends to increase Z_{Db} at higher frequencies. Counteracting this increase, the s^2 term of the denominator returns the net impedance response to a single-pole roll off like that attained with the R-C filter.

Thus the L-C filter offers two-pole roll off of supply-line noise and reduced low-frequency line impedance. However, at an intermediate frequency, the Z_{Db} impedance increases due to a new L-C resonance. From an energy perspective, the L_D inductor stores rather than dissipates the transient energy of supply-line noise, giving rise to this resonance. At most frequencies, the inductor simply releases its stored energy to the bypass capacitance and the amplifier in a more gradual manner than initially presented by a supply noise transient. This smoothing of the energy release protects the amplifier from supply-line noise by reducing the associated frequency content to a range where amplifier PSRR remains high. However, at the inevitable L-C resonance the Z_{Db} impedance rises and the filtering fails. As indicated by the Z_{Db} denominator, $1 + L_D C_B s^2$, the $L_D - C_B$ combination produces a double pole and a resonance at $f_{rDb} = 1/2\pi\sqrt{L_D C_B}$. This resonance increases rather than decreases the line impedance in the vicinity of f_{rDb}.

This resonance makes the simple L-C decoupling filter an uncertain alternative for op amp circuits having high noise gains. At resonance, energy stored in the inductor by supply-line transients circulates between the capacitor and the inductor until dissipated by some energy drain. In the ideal L-C case, this circulation continues indefinitely, sustaining an oscillation signal. In practice, parasitic resistances and the amplifier help absorb that circulating energy, but the amplifier's PSRR error and gain offer replenishment to support this oscillation. There supply-voltage transients or noise containing frequency components in the range of the f_{rDb} resonance can initiate this condition.

Adding a detuning resistor removes the resonance and the oscillation uncertainty but reduces the noise roll off. Shown in Fig. 3.13c, this alternative combines the R_D and L_D series impedances of the preceding two cases. The combination provides both the energy absorption of the R-C filter and the reduced low-frequency impedance of the L-C filter. For this R-L-C combination the filtering response becomes

$$V_P = \frac{[1 + (L_D/R_D)s]V_+}{1 + (L_D/R_D)s + L_D C_B s^2}$$

Here the s^2 term of the denominator initially continues the two-pole roll off advantage of the basic L-C case. However, the s term of the numerator indicates a response zero that returns this roll off to single-pole. Thus adding R_D to the L-C filter reduces the filter attenuation.

Still, R_D improves the decoupling effectiveness by detuning the line impedance resonance that otherwise adds gain to the parasitic PSRR feedback loop. Analysis of the Z_{Dc} impedance, as seen from the V_P terminal, yields

$$Z_{Dc} = \frac{L_D s}{1 + (L_D/R_D)s + L_D C_B s^2}$$

This expression repeats that of Z_{Db} except for the $(L_D/R_D)s$ term of the denominator. That term separates the circuit's two poles to detune the characteristic L-C resonance and remove the associated impedance rise.

3.5.2 Selecting the decoupling components

Two compromises guide the selection of the decoupling filter components. The first repeats the bypass capacitance selection described before, optimizing the value of C_B to maintain frequency stability. Design equations developed earlier guide this selection. The second compromise guides the selection of the series impedance to limit noise coupling without significantly degrading other performance.

The series impedance R_D, or the R_D and L_D combination, must be large enough to intercept high-frequency line transients, but not so large as to compromise the circuit's lower-frequency PSRR error signal. The power-supply decoupling reduces the supply-line noise produced by other circuitry but can increase the parasitic feedback produced by the amplifier itself. At lower frequencies, the bypass capacitors fail to roll off the line impedance, and adding the series impedance increases the supply-line impedance seen by the amplifier. Section 3.1 describes the noise-coupling effect of this impedance. The net effect upon circuit performance remains a function of multiple variables and discourages analysis. However, following the previous bypass capacitor selection guidelines and then setting $R_D \sim 5\ \Omega$ serves most amplifiers that supply output currents in the 10-mA range. This 5-Ω guideline, combined with the 1-Ω guideline for C_B, produces a filter with at least a 6:1 reduction of higher-frequency noise. Also, adding just 5 Ω generally retains the lower-frequency line impedance at an acceptable level. Amplifiers supplying greater output current require correspondingly lower values for R_D. For R-C filters, this R_D selection completes the decoupling design.

For R-L-C filters, the L_D selection remains. Adding the L_D inductor of Fig. 3.13c benefits performance as long as the inductance chosen forms a detuned resonant circuit with the C_B and R_D chosen before. Making L_D larger minimizes the line impedance at lower frequencies but potentially increases it at the L-C resonance. For optimum detuning of this resonance, the L_D impedance should equal R_D at the frequency of the potential resonance. From the discussion of Fig. 3.13b the undamped resonance would occur at the frequency $f_{rDb} =$

$1/2\pi\sqrt{L_D C_B}$. There the impedance of L_D equals $Z_{LD} = 2\pi L_D f_r = \sqrt{L_D/C_B}$, and setting this equal to R_D yields

$$L_D = R_D^2 C_B$$

Here the capacitance C_B represents the combined value of the bypass capacitors connected to the associated supply terminal.

References

1. J. Graeme, "Design Equations Help Optimize Supply Bypassing for Op Amps," *Electron. Des.*, June 24, 1996.
2. H. Ott, "Noise Reduction Techniques in Electronic Systems," 2d ed., Wiley, New York, 1988.
3. J. Graeme, "Feedback Plots Offer Insight into Operational Amplifiers," *EDN*, January 19, 1989, p. 131.
4. J. Graeme, "Feedback Models Reduce Op Amp Circuits to Voltage Dividers," *EDN*, June 20, 1991, p. 139.
5. J. Graeme, "Fast Op Amps Require More Than a Single-Capacitor Bypass," *Electronic Design*, November 18, 1996.

Chapter

4

Phase Compensation

The greatly varied applications of op amps routinely encounter response ringing or oscillation. Often in such cases, the general-purpose phase compensation included within the op amp proves to be insufficient for the application and additional phase compensation must be added. However, most op amps lack provision for external adjustment of their internal phase compensation. Then external adjustment of the amplifier feedback network provides the required compensation instead. This chapter describes eight feedback compensation methods, and one or more of these suits most any application. These compensation methods modify either the net open-loop response of the circuit or the circuit's 1/ß response. However, before applying any of these methods, one must ensure that the power-supply bypass measures of Chap. 3 have been applied correctly. Frequently, inadequate supply bypass underlies an encountered instability.

The need for additional phase compensation arises from a variety of circuit conditions. Most frequently, this compensation adapts the op amp circuit to capacitance loading or amplifier input capacitance. In other cases, external phase compensation permits the use of lightly compensated, higher-speed amplifiers in low-gain circuits. In still other cases, the external compensation accommodates the two-pole, open-loop responses of composite amplifiers. Two of the compensation methods described here specifically address capacitance loading requirements through modification of the loaded circuit's open-loop gain response.[1] The first of these requires empirical component selection, but it also offers a unique filtering action. The second method also modifies the circuit's open-loop response and removes much of the previous empirical component selection. However, this second method can restrict the output voltage range.

The remaining phase compensation methods modify a circuit's 1/ß response instead of the open-loop response. The first four address the effects of op amp input capacitance, with three of these compensating the basic inverting, noninverting, and differential op amp configurations. The fourth adjusts the noninverting circuit compensation for higher source resistance cases. Amplifier input capacitance affects the four circuit cases differently, requiring specialized phase compensation. In each case, standard design equations define the compensation for the more common resistive-feedback applications. The final two phase compensation methods address all phase compensation requirements by again modifying the circuit's 1/ß response. These methods offer an opportunity for increased slew rate through the use of lighter-compensated op amps. The first method modifies the circuit's negative feedback, and this especially suits integrator and differential input configurations. The second method adds positive feedback for greater bandwidth from voltage followers.

4.1 Phase Compensation for Capacitance Loading Effects

Capacitive load drive produces the most common requirement for added phase compensation of op amps. Such loads degrade stability by introducing a second pole in the feedback path. Two phase compensation methods in this section address this condition specifically. However, the phase compensation methods described in Sec. 4.3 also satisfy this requirement. The first method described here decouples the amplifier from the load capacitance by providing a bypass feedback path for the amplifier. The second method applies pole-zero phase compensation and provides greater analytical predictability.

4.1.1 Capacitance loading and frequency stability

Figure 4.1 demonstrates the circuit effect of capacitance loading on an op amp. Load C_L reacts with the amplifier's open-loop output resistance R_o, producing an added pole in the feedback path. This pole increases the feedback phase shift and results in response ringing or even oscillation. To quantify the response degradation, Fig. 4.2 shows the amplifier's open-loop response A_{OL} and the circuit's inverse feedback factor 1/ß. The loaded and unloaded A_{OL} curves of the plot illustrate the capacitance loading effect upon the amplifier response. Evaluation of these two curves shows that the loaded A_{OL} response reduces circuit bandwidth and compromises frequency stability.

As described in Chap. 1, the circuit bandwidth depends directly upon the intercept frequency of the 1/ß curve with the A_{OL} response. In the figure, capacitance loading reduces the intercept frequency

Phase Compensation

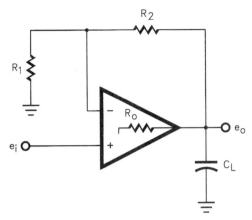

Figure 4.1 Capacitance loading of op amp output resistance introduces a low-pass filter in the feedback path.

from f_i to f_i'. At the intercept, the feedback demand for gain $1/ß$ and the available amplifier gain A_{OL} cross. There $A_{OL} = 1/ß$, and the available gain just equals the feedback demand. Beyond this frequency, the amplifier lacks sufficient gain for continuance of the full circuit response. Thus the circuit's closed-loop response rolls off, defining a bandwidth limit equal to the intercept frequency. As illustrated, capacitance loading reduces this limit by a factor of f_i/f_i' to BW $= f_i'$.

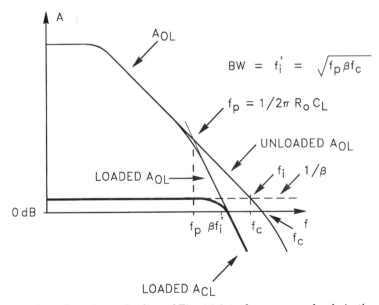

Figure 4.2 Capacitance loading of Fig. 4.1 introduces a second pole in the A_{OL} response, producing a -40-dB per decade slope at the $1/ß$ intercept.

The circuit's frequency stability also depends upon the intercept frequency, but in a less direct manner. As described in Sec. 1.3, the slope relationship of the 1/ß intercept with the A_{OL} response predicts the circuit's stability conditions. At any frequency, the difference in slopes of the two curves predicts the net phase shift in the feedback loop. However, this phase shift only becomes important at the 1/ß intercept, where the circuit's loop gain equals unity and could support oscillation. To preserve frequency stability, this slope difference at the intercept, or rate of closure, must be limited. For Fig. 4.2, the 1/ß curve presents zero slope, leaving the rate of closure a function of the A_{OL} slope alone. In the unloaded case, A_{OL} follows a single-pole roll off at the f_i intercept, producing a rate of closure of 20 dB per decade, and this rate predicts a stable 90° phase shift in the loop at this critical intercept. However, in the loaded case, C_L alters the A_{OL} response, introducing a second pole at $f_p = 1/2\pi R_o C_L$. The resulting two-pole or −40-dB per decade A_{OL} slope signals a phase shift that eventually reaches 180°. Oscillation results if this 180° condition occurs at or before the new f_i' intercept with the 1/ß curve. Even if oscillation does not occur, this second pole increases response overshoot and ringing, introduces gain peaking, and limits bandwidth. As will be described, added phase compensation removes the overshoot, ringing, and peaking, but it fails to restore the bandwidth.

The circuit's phase margin more precisely quantifies the phase shift indicated by the rate-of-closure evaluation. At the 1/ß intercept, 180° of feedback phase shift produces oscillation. There the phase margin is the amount of phase separating the feedback phase shift from the critical 180°. Typically, a phase margin between 45 and 60° limits overshoot to about 30%, minimizes ringing, and limits gain peaking to around 3 dB. For the unloaded case of Fig. 4.2, the single-pole A_{OL} response produces a phase shift $\phi_i = 90°$ at the f_i intercept, resulting in a phase margin of $\phi_m = 180° - \phi_i = 90°$. The second pole of the loaded case increases the net phase shift at the f_i' intercept to $\phi_i' = 90° + \tan^{-1}(f_i'/f_p)$. This reduces the phase margin to $\phi_m' = 180° - \phi_i' = 90° - \tan^{-1}(f_i'/f_p)$. Thus capacitance loading reduces the phase margin by the amount of $\tan^{-1}(f_i'/f_p)$.

To quantify BW and ϕ_m' for the loaded case, the frequency f_i' must first be defined in terms of known quantities. The frequencies f_p and f_i serve as reference points for the associated analysis. Calculation or, better yet, measurement readily determines f_p, and prior analysis results define f_i. This intercept results from a flat 1/ß curve and a single-pole A_{OL} response. In such a case, $f_i = ßf_c$, as described in Sec. 1.2.3, where f_c is the unity-gain crossover frequency of the unloaded amplifier.

A graphical analysis relates f_i' to f_p and $ßf_c$ by exploiting geometric relationships between the curves in Fig. 4.2. Note that the dashed vertical line indicating f_p forms right triangles bounded by the dashed 1/ß curve and the two open-loop responses. The hypotenuse of the loaded triangle has a two-pole slope, twice that of the unloaded triangle, making the base length of the loaded triangle one-half that of the unloaded one. Given the logarithmic nature of the frequency axis, this base length relationship translates to $\log f_i' - \log f_p = 0.5(\log ßf_c - \log f_p)$. Solving this expression defines f_i' as the geometric mean of f_p and $ßf_c$, or $f_i' = \sqrt{f_p ßf_c}$. Combining this result with the previous BW and ϕ_m' expressions defines these characteristics in terms of known quantities. Then, for a capacitively loaded op amp,

$$BW = f_i' = \sqrt{f_p ßf_c}$$

$$\phi_m' = 90° - \tan^{-1} \sqrt{ßf_c/f_p}$$

where $f_p = 1/2\pi R_o C_L$. A condition of $\phi_m' < 45°$ typically signals the need for additional phase compensation.

4.1.2 Decoupling the capacitance load

A first phase compensation method decouples the amplifier from the capacitance load, as shown in Fig. 4.3. There resistor R_C decouples R_o from C_L while C_C provides a bypass feedback path for the amplifier. Given the appropriate design, this decoupling method permits stable drive of any amount of capacitance load. Although shown with the noninverting configuration, the compensation technique serves

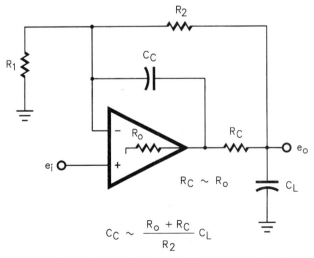

Figure 4.3 Decoupling resistor R_C and bypass capacitor C_C isolate an op amp from a capacitance load.

almost any op amp configuration. However, the use of this technique with differential input connections degrades common-mode rejection, as discussed in Sec. 4.3.7. Also, for use with a voltage follower, this technique requires the independent addition of resistor R_2 as part of the phase compensation.

The decoupling phase compensation provides a bypass feedback loop that overrides the primary feedback loop at higher frequencies. With two feedback paths, the control of the op amp depends upon which path prevails in supplying the feedback current. Feedback supplies this current to R_1, making the voltage on this resistor follow e_i. Either R_2 or C_C supplies this current, depending upon the signal frequency. At low frequencies, C_C presents a large impedance and R_2 dominates the feedback current supply. There the primary feedback loop controls the circuit response and produces the familiar $e_o = (1 + R_2/R_1)e_i$. At high frequencies, C_C prevails and supplies R_1 through the bypass feedback path, reducing e_o to zero. Between these frequency extremes, the feedback control of the amplifier makes a transition between the two feedback paths.

Figure 4.4 shows the relevant response curves. There three different high-frequency responses represent A_{OL} for the uncompensated case and for the primary and bypass loops of the compensated case. Capacitance loading still produces a two-pole roll off for the primary loop, but the bypass loop isolates the amplifier from this roll off. The

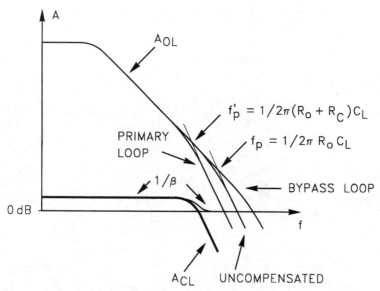

Figure 4.4 Phase compensation of Fig. 4.3 retains single-pole A_{OL} response in a bypass feedback loop, avoiding two-pole response in the primary loop.

bypass loop retains stable feedback conditions with an A_{OL} curve largely unaffected by the capacitance load. At first, the addition of R_C and C_C would seem to aggravate the stability problem. For the primary-loop response, the addition of R_C moves the capacitance-loading pole to an even lower frequency f_p'. Also, the addition of C_C causes the 1/ß curve to drop to the unity-gain axis, and this 1/ß = 1 condition represents the most demanding case for feedback stability. At the unity-gain axis, the 1/ß curve intercepts the primary-loop response in a region of two-pole roll off, suggesting poor frequency stability or even oscillation.

Fortunately, this intercept does not reflect the op amp's feedback conditions. The A_{OL} curve of the primary loop only represents the response seen from the circuit output. For stability considerations, the relevant A_{OL} curve represents the response seen from the amplifier, rather than circuit, output. At higher frequencies, direct feedback from the amplifier output controls the amplifier through the C_C bypass, and the bypass A_{OL} response remains virtually unaffected by the capacitance load due to the decoupling provided by R_C. For the op amp's feedback, the bypass and decoupling preserve the original amplifier response as represented by the bypass-loop A_{OL} curve. This curve presents a reduced slope at its intersection with 1/ß, restoring stability to the capacitively loaded amplifier.

4.1.3 Selecting the decoupling components

The complex nature of an op amp's open-loop output impedance complicates the component selection for this decoupling phase compensation. This output impedance reacts with the load capacitance, producing the pole to be countered by the phase compensation. For the resistive output impedance R_o modeled, simple analysis defines the pole frequency. However, the actual output impedance of an op amp displays a complex frequency behavior. A typical amplifier produces an open-loop output impedance starting at around 100 Ω to 1 kΩ at dc. From there this impedance drops dramatically to 10–50 Ω at higher frequencies. At even higher frequencies, this impedance may rise again. Also, the output impedance varies with the instantaneous level of the amplifier's output current.

The task of modeling this output impedance behavior exceeds that of the empirical alternative. With that alternative, circuit analysis using the simple R_o model first provides a rough guide to values for R_C and C_C. Then empirical testing guides an adjustment of C_C to accommodate the true output impedance under varying output load conditions. Note that circuit simulation with op amp models encounters this same constraint since SPICE op amp models typically model high- and low-frequency output impedances with simple resistors. Thus decoupling phase compensation selected through simulation should also be fine-tuned through empirical tests.

Guidelines aid in the initial selection of both R_C and C_C. First, choose R_C to be roughly equal to high-frequency output impedance of the amplifier, typically 10–50 Ω. To determine this value, open-loop response measurement defines f_p, and then $R_o = 1/2\pi C_L f_p$ approximates this impedance. Intuition might suggest even higher resistance for R_C to increase the decoupling from load C_L. However, R_C should not be made arbitrarily large because of the voltage drop produced on this resistance by the amplifier's output current. This voltage drop does remain inside the primary feedback loop, avoiding gain error, but the drop detracts from the output voltage range. Also, larger values of R_C restrict the circuit's bandwidth, as described later. Fortunately, even small values of R_C reduce the phase shift developed across R_o by C_L dramatically, and only this phase shift alters the bypass drive of C_C to affect the final circuit stability.

After selecting R_C, a second guideline defines the initial value for C_C. Select a C_C value that transfers feedback control to the bypass path at the frequency f_p'. At this frequency, the second pole of the primary loop would begin to compromise stability. However, transferring feedback control to the bypass loop removes the second-pole effects from the new controlling feedback to preserve stability. To calculate the initial C_C value, approximate f_p' with the relationship $f_p' = 1/2\pi(R_o + R_C)C_L$, where R_o equals the value determined in the R_C selection.

The next calculation sets the break frequency of C_C with the $R_C + R_2$ feedback path at the f_p' value. For simplicity, this calculation ignores the presence of C_L, but a simple evaluation justifies this simplification for the purposes of this design guideline. Capacitance C_L breaks with R_C, shunting the feedback of the $R_C + R_2$ path, and this break occurs at approximately $1/2\pi R_C C_L \approx f_p$ for $R_C \approx R_o$. However, setting the C_C break frequency at f_p' places this latter break at about one-half the frequency f_p. To demonstrate this, consider the prescribed $R_C \approx R_o$, which makes $f_p' = 1/2\pi(R_o + R_C)C_L \approx 1/2\pi(2R_o)C_L \approx f_p/2$. At $f_p/2$, or one-half of that break frequency, the voltage divider formed by R_C and C_L produces a divider ratio of $1/\sqrt{1.25} = 0.9$. Thus the $R_C + R_2$ feedback path continues to transmit 90% of its original signal at the frequency f_p' considered here.

Ignoring C_L and its 10% attenuation effect greatly simplifies the C_C selection and still provides adequate accuracy for the selection guideline. Then setting $1/2\pi(R_C + R_2)C_C = f_p' = 1/2\pi(R_o + R_C)C_L$ and solving for C_C defines this capacitance's initial design equation. Assuming $R_2 \gg R_o$, the initial compensation becomes

$$C_C \approx \frac{R_o + R_C}{R_2} C_L \quad \text{and} \quad R_C \approx R_o$$

For very large capacitance loads, this initial C_C becomes too conservative, but the next step corrects for this.

Following the initial selection of R_C and C_C, empirical testing guides the final compensation adjustment. With the compensation components in place, adjust the value of C_C while observing the circuit's square-wave response. Adjust C_C to limit overshoot to 30% or less and to minimize response ringing. Make this adjustment with tests covering the full range of C_L, load resistance, and output current values expected in the application. Fortunately, the compensated circuit response presents a low sensitivity to the resistance and capacitance values. Degraded, but stable, response remains for around a 10:1 range above or below the design center. Thus the 2:1 variations in R_o due to manufacturing tolerances do not greatly affect the resulting circuit stability.

In the C_C adjustment process, conservative design practice might suggest increasing capacitance just for the absolute assurance of stability. However, excessive C_C values unnecessarily restrict the circuit's bandwidth. The C_C bypass diverts signal away form the circuit output, causing the output response to roll off at the transition frequency of the circuit's two feedback loops. The earlier derivation of the C_C design equation approximates this roll-off frequency by the break of C_C with the $R_C + R_2$ feedback path. This break frequency approximates the circuit bandwidth by

$$\text{BW} \approx 1/2\pi(R_C + R_2)C_C$$

Thus bandwidth considerations favor minimizing C_C. In compromise, C_C should be made large enough to preserve stability, but not so large as to limit bandwidth.

4.1.4 Filtering provided by decoupling

The phase compensation of Fig. 4.3 also provides a unique filtering action, which rejects amplifier noise better than most op amp filter circuits.[2] Without the phase compensation (Fig. 4.3) and the analogous inverting configuration, amplify the input voltage noise of the op amp by a gain of $1/ß = 1 + R_2/R_1$. To filter out high-frequency noise, it is common to bypass R_2 with a capacitor. However, this bypass only removes the R_2/R_1 portion of the gain received by the op amp noise. The amplifier's high-frequency noise continues to receive a gain of $1/ß = 1$ up to the open-loop roll off of the op amp. This same condition holds for most op amp connections, including active filter circuits. In low-frequency applications, the continued noise gain frequently dominates noise performance.

With the decoupling phase compensation, the filter formed by R_C and C_L prevents the continuation of this noise gain. At higher frequencies, the C_C feedback controls the op amp itself with a unity feedback factor. Thus the amplifier's input noise still receives unity gain up to the op amp output. However, between this output and the actual circuit output, R_C and C_L form a low-pass filter, and this filter shunts the high-frequency amplifier noise to ground. For filter applications, C_L becomes an intentionally added element in the circuit design.

4.1.5 Pole-zero compensation for capacitance loading

A second method of external phase compensation removes most of the empirical component selection described. However, this method does reduce the circuit's output voltage range when supplying significant output currents. Shown in Fig. 4.5, this second method introduces a paralleled resistor and capacitor in series with the amplifier output but inside the feedback loop. This combination produces a pole-zero compensation for a capacitively loaded amplifier. As illustrated, the circuit is a voltage follower, but this compensation method applies to all op amp configurations. The follower example illustrates the phase compensation effect without the unrelated feedback effects of other configurations.

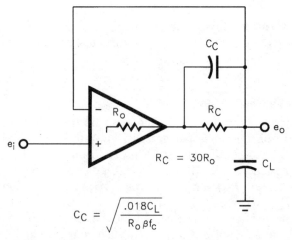

Figure 4.5 Pole-zero phase compensation for capacitance loading adds control of pole frequency through R_C and adds a zero through C_C bypass of R_C.

With this phase compensation, the capacitance load that creates the problem becomes part of the solution. As before, load C_L reacts with the circuit's open-loop output resistance, creating a second response pole. In the uncompensated case, C_L reacts with R_o, compromising frequency stability, as described with Fig. 4.2. The addition of R_C increases the circuit's open-loop output resistance but permits greater control over the second pole's effects and makes C_L part of the phase compensation. The added resistor moves the second pole back in frequency, creating the opportunity for a response zero. To produce this stabilizing zero, capacitor C_C bypasses R_C to counteract the resistor's effect at higher frequencies.

Figure 4.6 shows the effect of this phase compensation upon A_{OL}. There the uncompensated and compensated A_{OL} curves present different slopes at their 1/ß intercepts. In this voltage follower example, 1/ß = 1, and the corresponding curve follows the 0-dB axis. The uncompensated A_{OL} curve intercepts this axis at f_i with a two-pole roll off. Phase compensation reduces this roll-off slope to single-pole for an intercept at f_i'. Classic pole-zero compensation produces this compensated result. Adding R_C first moves the f_p pole back to f_p', and the resulting A_{OL} response remains free to be redirected anywhere within the bounds of the uncompensated response. Then the addition of C_C redirects the

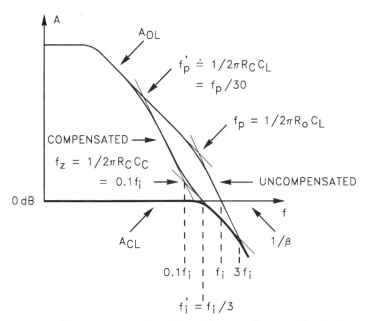

Figure 4.6 Phase compensation of Fig. 4.5 moves the capacitive loading pole back in frequency to permit a response zero that restores a stable intercept.

compensated response through a response zero as the C_C bypass of R_C temporarily removes the effect of f_p', restoring a -20-dB per decade response slope. This reduced slope continues until the compensated response reaches the boundary imposed by the uncompensated response. At that point, a renewed shunting of R_o by C_L rejoins the two A_{OL} responses. For good stability, the phase compensation selection places the region of reduced A_{OL} slope to span the 1/ß intercept.

Many different pole-zero combinations provide this placement, and different combinations actually optimize various op amp configurations. This suggests different compensation selection methods for each configuration. Fortunately, the circuit stability remains fairly insensitive to the pole-zero combination, and one selection method adequately serves all configurations. For this selection, phase margin analysis at the f_i' intercept defines a universal pole-zero combination. As described before, the phase margin equals 180° minus the feedback phase shift developed at the 1/ß intercept. In this case, the feedback phase shift at f_i' depends upon the frequency duration of the compensation's -20-dB per decade region. For 90° of phase margin, this region would be set to span about one decade of frequency both before and after f_i'. Experience shows the decade after f_i' to be wise because, there, secondary amplifier poles add even more phase shift. However, a decade span before f_i' unnecessarily restricts bandwidth and a one-half-decade range suffices. Antilog conversion equates this one-half decade to about a factor of 3 in frequency. Thus the total span of reduced A_{OL} slope should cover a 30:1 frequency range, spaced about f_i', as defined by the preceding text.

The compensated response shown in Fig. 4.6 closely approximates these conditions. Relating the figure's various break frequencies to f_i introduces this frequency as a common factor and demonstrates the two span limits of the reduced slope around f_i'. First, the compensation shown conveniently sets $f_i' = f_i/3$. Graphical analysis shows that this setting produces a -20-dB per decade response up to $3f_i$, where the boundary imposed by the uncompensated response returns the slope to -40 dB per decade. There the span termination at $3f_i$ corresponds to a frequency of $9f_i'$. Thus above f_i', the reduced response slope covers a frequency range of 9:1, or almost the decade described. Prior to this intercept, the reduced slope extends back to the limit set by f_z. Making $f_z = 0.1f_i$ results in $f_z = 0.3f_i' \approx f_i'/3$. Thus the overall -20-dB per decade span extends from $0.1f_i$ to $3f_i$ for the desired 30:1 frequency range.

4.1.6 Selecting the pole-zero compensation components

Component selection follows from the preceding conditions and the measurement of the frequency f_p. First, the relationship between frequencies f_p and f_p' defines the resistor R_C through graphical analysis of the Fig. 4.6 response curves. There the straight-line extensions of the compensated and uncompensated responses form a parallelogram. The equal lengths of opposite sides of a parallelogram establish that the distance between f_p and f_p' equals that between $3f_i$ and $0.1f_i'$, making $f_p = 30 f_p'$. Then simply setting $R_C \approx 30 R_o$ permits the desired 30:1 span of the reduced slope. Note that the higher R_C value here reduces the output voltage range due to the voltage drop developed on this resistor by output current.

However, $R_C \approx 30 R_o$ also makes $R_C \gg R_o$, removing much of this phase compensation's sensitivity to the complex nature of the actual amplifier output impedance. This eliminates the empirical fine-tuning described for the previous compensation solution. Only the measurement of f_p, to determine R_o, remains as an empirical task in the phase compensation selection. However, the removal of empirical fine-tuning increases the requirement for accurate knowledge of the effective R_o value. Op amp data sheets do not reflect the specific R_o value that applies at a given f_p frequency of interest. Empirical measurement of f_p defines the appropriate value through $R_o = 1/2\pi f_p C_L$. Then $R_C = 30 R_o$ defines the compensation resistance.

Having selected R_C, f_z now defines C_C. Compensation conditions selected before set $f_z = 1/2\pi R_C C_C = 0.1 f_i'$. This makes $C_C = 5/\pi R_C f_i$ and requires an expression for f_i to complete the solution. Previously, the analysis of Fig. 4.2 defined this frequency as $\sqrt{f_i} = f_p \beta f_c$, where $f_p = 1/2\pi R_o C_L$. Combining this result with the previous C_C expression reduces the expression to a function of known quantities. For the phase compensation of Fig. 4.5,

$$C_C = \sqrt{\frac{0.018 C_L}{R_o \beta f_c}} \quad \text{for } R_C = 30 R_o$$

These values for R_C and C_C typically retain a degraded but stable response for a 10:1 range for the value of the load capacitance C_L.

With this compensation, the $1/\beta$ intercept again defines the circuit's bandwidth. As described, this intercept occurs at $f_i' = f_i/3$. Again borrowing from the analysis in Fig. 4.2, $f_i = \sqrt{f_p \beta f_c}$, where $f_p = 1/2\pi R_o C_L$. Combining these equations defines the bandwidth in Fig. 4.5 as

$$BW = f_i' = \sqrt{\frac{0.018 \beta f_c}{R_o C_L}}$$

For the voltage follower example here, the $1/\beta$ curve follows the 0-dB or unity-gain axis to define the critical intercept. In other op amp configurations, the $1/\beta$ curve shifts upward, moving the bandwidth-defining intercept. The preceding equations for C_C and BW automatically adjust for this difference through their β terms.

Note that this pole-zero compensation compromises the settling time. The added pole and zero produce a frequency doublet, notorious for its poor settling time.[3] Where this becomes important, the later phase compensation techniques (see Fig. 4.13 or Fig 4.17) offer better settling performance. As presented there, those circuits also produce an added pole and zero in the feedback path. However, those circuits permit elimination of the compensation capacitor C_C to remove the pole-zero pair.

4.2 Phase Compensation for Input Capacitance Effects

Op amp input capacitance produces frequent, and often surprising, performance disturbances through degraded frequency stability. In those cases, the input capacitance forms a response pole with the feedback network, introducing additional feedback phase shift. Higher-frequency op amps exhibit exceptional vulnerability to this capacitance effect. For noninverting op amp connections, this capacitance also produces gain peaking by bypassing one of the gain-setting feedback elements. This gain peaking also jeopardizes common-mode rejection in differential input op amp circuits.

Four phase compensation methods counteract this effect, with the compensation choice depending upon the circuit configuration and the signal source conditions.[4] Three methods compensate the basic inverting, noninverting, and differential op amp configurations. The fourth adjusts the noninverting circuit compensation for higher source resistances. In each case, Bode analysis yields standard design equations that define the compensation components and the resulting bandwidth. The analyses also define mathematical tests that determine whether a circuit requires phase compensation and, if so, which kind. These analyses focus upon the resistive feedback most commonly found in op amp applications. They also focus upon the input capacitances of the op amp, but the results extend to other capacitance effects as well. Other circuit capacitances at the amplifier input simply add to those of the op amp, both in the circuit and in the design equations. Added input capacitance most often occurs in photodiode amplifier[5] applications.

The simple phase compensation first described for the inverting configuration adequately restores stability for any op amp configuration. However, bandwidth considerations differentiate the phase compensation most suitable for the three basic op amp configurations. Inverting configurations achieve stable operation with maximum bandwidth from a minimum-value compensation capacitance. Surprisingly, noninverting configurations sometimes achieve greater bandwidth from larger capacitance values. There the phase compensation also serves to counteract gain peaking. This gain peaking afflicts differential input configurations as well, and this configuration requires two balancing compensation capacitances to preserve common-mode rejection. Alternately, this balanced compensation also preserves bandwidth for noninverting configurations when driven from higher source resistances.

4.2.1 Input capacitance and frequency stability

Figure 4.7 illustrates the fundamental problem produced by capacitance at the inverting input of an op amp. There the resistive-feedback inverting case demonstrates the problem with the components of the amplifier's input capacitance drawn outside the amplifier. There

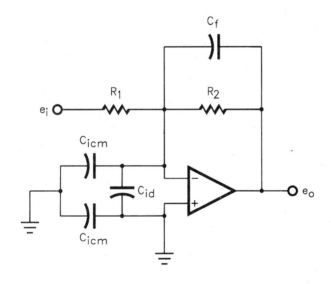

For $f_z < \beta_0 f_c$, add

$$C_f = (C_c/2)\left[1 + \sqrt{1 + 4C_i/C_c}\right]$$

where $C_c = 1/2\pi R_2 f_c$

and $C_i = C_{id} + C_{icm}$

Figure 4.7 In inverting cases, feedback capacitance C_f produces a feedback zero, compensating for the pole introduced by input capacitance.

C_{id} represents the amplifier's differential input capacitance between the two amplifier inputs, and the two C_{icm} components represent the common-mode input capacitances from each input to ground. In this case, two of the three capacitances shunt the feedback network. Feedback analysis first demonstrates the stability effect of these capacitances, beginning with the feedback factor ß. As described in Chap. 1, the feedback factor is the fraction of the amplifier output signal fed back to the amplifier's input. Superposition grounding of e_i clarifies the determination of ß for the inverting circuit considered here. This grounding places R_1 in parallel with C_{id} and the upper C_{icm} component, making the relevant input capacitance $C_i = C_{id} + C_{icm}$. As will be seen, adding the C_f shown in parallel with R_2 phase compensates the circuit to counteract the stability effects of C_i.

For the inverting op amp connection here, the lower C_{icm} component in Fig. 4.7 produces no effect. Both sides of this capacitance connect to common, short-circuiting that capacitance. In other configurations, the lower C_{icm} alters the signal supplied to the op amp's noninverting input. However, this signal alteration does not affect the feedback signal delivered to the amplifier's inverting input. Thus the feedback factor and stability results described here for the inverting configuration still hold for all other op amp configurations.

To examine the initial stability effect of $C_i = C_{id} + C_{icm}$, consider the uncompensated case where $C_f = 0$. Then the feedback voltage divider that defines ß consists of R_2 and the parallel combination of R_1 and C_i. The associated voltage divider ratio defines ß as

$$ß = \frac{ß_0}{1 + (R_1 || R_2)C_i s}$$

where $ß_0 = R_1/(R_1 + R_2)$, $C_i = C_{id} + C_{icm}$, and $C_f = 0$. Thus the op amp input capacitance introduces a feedback pole at $f = 1/2\pi(R_1 || R_2)C_i$.

This feedback pole shunts the feedback signal supplied to the amplifier's inverting input, potentially degrading stability. The significance of this pole depends upon the resistance levels of the feedback network and the bandwidth of the op amp. Lower resistance networks form a pole with C_i well beyond the reaches of the amplifier's open-loop response, and there the associated phase shift has little effect upon circuit stability. However, higher resistance networks or wider-band op amps include this pole's phase shift within the amplifier's response range. Then phase compensation must be added to restore good ac response characteristics.

As shown in Fig. 4.7, the capacitive bypass of R_2 compensates for the effect of C_i. There the C_f bypass introduces a zero in the feedback path to counteract the pole produced by C_i. Figure 4.8 demonstrates the stabilizing effect of C_f through the circuit's 1/ß response and open-

Phase Compensation

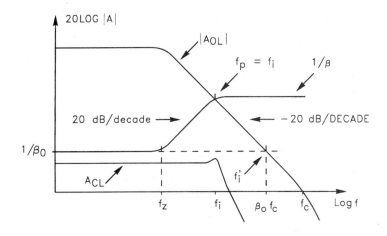

$$\beta_0 = \frac{R_1}{R_1 + R_2} \qquad f_z = 1/2\pi(R_1 \| R_2)(C_i + C_f)$$

$$f_p = f_i = \sqrt{\frac{f_c}{2\pi R_2(C_i + C_f)}} \qquad BW = 1.4 f_i$$

Figure 4.8 Phase compensation of Fig. 4.7 restores stable intercept by leveling off the 1/ß curve following a rise produced by input capacitance.

loop gain curve. As described in Chap. 1, these curves predict an op amp circuit's stability conditions through their intercept. Analysis of the circuit defines the 1/ß response as

$$1/\text{ß} = \frac{1 + (R_1 \| R_2)(C_i + C_f)s}{1 + R_2 C_f s}\left(1/\text{ß}_0\right)$$

where $\text{ß}_0 = R_1/(R_1 + R_2)$. This response begins at a low-frequency value of $1/\text{ß}_0$ and then experiences a response zero at $f_z = 1/2\pi(R_1 \| R_2)(C_i + C_f)$, followed by a pole at $f_p = 1/2\pi R_2 C_f$.

In Fig. 4.8 the zero at f_z makes the 1/ß response rise with a slope of 20 dB per decade toward an A_{OL} curve having a slope of -20 dB per decade. These opposite slopes produce the potential for a 40-dB per decade rate of closure, or slope difference, at the f_i intercept. That rate predicts as much as 180° of feedback phase shift, warning of oscillation. However, to produce the 40-dB per decade closure, f_z must occur within the response range of the amplifier. In the figure, this requires that f_z occur before the 1/ß curve crosses the A_{OL} curve. If this does not occur, the relevant 1/ß curve simply continues along the path denoted by the level dashed line and makes a stable intercept with A_{OL}. As illustrated, this intercept occurs at the frequency f_i', defining the boundary for f_z influence upon the amplifier response.

4.2.2 Testing for input capacitance effects

The capacitance at an op amp's inverting input may or may not affect stability conditions. As described in the preceding section, the effect of this capacitance only matters when the associated frequency f_z drops below f_i'. So far, the frequency f_i' remains an unknown, but simple analysis quantifies it in terms of known circuit parameters. At their f_i' intersection, the A_{OL} and dashed $1/ß_0$ curves of Fig. 4.8 occupy the same point, making $A_{OL} = 1/ß_0$. Also, in the region of this intersection, the typical A_{OL} response of an op amp exhibits a single-pole roll off, making $A_{OL} = f/f_c$, where f_c is the unity-gain crossover frequency of the amplifier. Thus at f_i', $A_{OL} = f_i'/f_c = 1/ß_0$, and solving for f_i' defines the intersection frequency as $f_i' = ß_0 f_c$. Then for $f_z < f_i' = ß_0 f_c$, the amplifier's input capacitance degrades the circuit stability. For $f_z > ß_0 f_c$, the amplifier lacks the gain-bandwidth product needed to support a resulting oscillation. Thus comparison of f_z and $ß_0 f_c$ determines whether or not the circuit requires a C_f compensation capacitance.

For this test, assume the uncompensated case with $C_f = 0$ when calculating f_z. With $f_z = 1/2\pi(R_1 || R_2)C_i$, this test depends upon the resistance level $R_1 || R_2$ and not upon the resistance of R_1 or R_2 alone. This fact permits an intuitive anticipation of a C_f compensation requirement based upon a circuit's resistance levels and closed-loop gain. Making either feedback resistor a lower value moves f_z to a higher frequency, reducing the probability of a C_f requirement. As a result, general-purpose op amps seldom require the C_f compensation when either R_1 or R_2 remains below about 10 kΩ. For broadband op amps, this resistance level drops to around 500 Ω. This resistance-level guideline also translates to a closed-loop gain guideline. With $A_{CL} = -R_2/R_1$, an $A_{CL} \gg 1$ typically involves a smaller resistance for R_1. Similarly, an $A_{CL} \ll 1$ typically involves a smaller resistance for R_2. In both cases, C_f is seldom required. However, $10 > A_{CL} > 0.1$ prompts an evaluation of resistance levels and then, perhaps, the f_z versus $ß_0 f_c$ test.

When required, the addition of C_f restores stability, as shown by the $1/ß$ curve in Fig. 4.8. Adding C_f levels off the $1/ß$ response with the pole at f_p. There the feedback zero produced by C_f becomes a pole in the $1/ß$ response, returning the response slope to zero. Depending upon the f_p location, the $1/ß$ leveling reduces the rate of closure to as low as 20 dB per decade, providing as much as 90° of phase correction and assured stability. For a given phase compensation, the residual feedback phase shift lies somewhere between the 180° worst case and the 90° best case, and phase compensation placement of f_p determines the actual result. Stability and bandwidth considerations guide this placement, with different results for different op amp configurations.

4.2.3 Compensating the inverting configuration

In Fig. 4.8, placing f_p well before the f_i intercept would reduce the feedback phase shift to 90° for absolute assurance of stability. However, such a setting restricts the inverting circuit's bandwidth excessively. In Fig. 4.7, the C_f bypass of R_2 also shunts the signal developed on this resistor, converting the circuit's gain from $-R_2/R_1$ to $-Z_2/R_1$. There $Z_2 = R_2/(1 + R_2C_f s)$, and the shunting by C_f limits the circuit's bandwidth to $f_p = 1/2\pi R_2 C_f$. Alternately, minimizing C_f extends this limit, but only up to a second bandwidth limit imposed at f_i. At this intercept, the feedback demand for amplifier gain $1/\beta$ equals the available open-loop gain A_{OL}. Beyond the intercept, the demand exceeds the available gain, forcing the closed-loop response to roll off. A good compromise makes the two bandwidth limits coincident, as illustrated in Fig. 4.8. Placing f_p at the critical intercept f_i introduces 45° of phase correction at this intercept and extends bandwidth to the f_i limit. Then the feedback phase shift at f_i reduces from 180 to 135°, leaving 45° of phase margin. This amount of phase margin permits a small closed-loop response peak that extends the -3-dB bandwidth to BW $= 1.4 f_i$.

To quantify f_i and C_f, a geometric analysis of the response curves first defines the frequency f_i. Note the triangle formed by the $1/\beta$ rise, the A_{OL} fall, and the dashed extension of the $1/\beta_0$ curve. Equal and opposite slopes produce the rising and falling sides of this triangle, placing the triangle peak over the midpoint of the base. Thus f_i, at the peak, lies midway between the f_z and $\beta_0 f_c$ endpoints of the base. The logarithmic nature of the frequency axis makes this midpoint log $f_i = (\log f_z + \log \beta_0 f_c)/2$. Solving for f_i defines the intercept frequency as the geometric mean of the frequencies defining the triangle base, or

$$f_i = \sqrt{f_z \beta_0 f_c} = \sqrt{\frac{f_c}{2\pi R_2 (C_i + C_f)}}$$

Setting $f_p = 1/2\pi R_2 C_f = f_i$ and solving yields the design equation for C_f. A fictitious capacitance, $C_c = 1/2\pi R_2 f_c$, simplifies the result for better intuitive evaluation. This C_c represents the value of the capacitance that would break with resistance R_2 at the amplifier's crossover frequency f_c. Including this C_c in the C_f solution yields

$$C_f = \frac{C_c}{2}\left(1 + \sqrt{1 + \frac{4C_i}{C_c}}\right)$$

where $C_c = 1/2\pi R_2 f_c$ and $C_i = C_{id} + C_{icm}$.

In many cases, solving this expression yields subpicofarad values for C_f. Then the parasitic effects of a practical circuit generally introduce this level of capacitance automatically, removing the need for C_f. In other cases, solving the C_f equation yields realistic values but does not necessarily indicate the need for that capacitance. As described before, a circuit only requires this capacitance when $f_z < ß_0 f_c$.

4.2.4 Compensating the noninverting gain peaking

The preceding phase compensation also provides adequate compensation for the noninverting configuration. However, this configuration often achieves greater bandwidth when increased phase compensation counteracts gain peaking as well. This opportunity occurs when even greater feedback resistance levels introduce a new input capacitance effect. There op amp input capacitance produces gain peaking by bypassing one of the gain-setting feedback resistors. A counteracting increase in the value of C_f removes this peaking, extending the bandwidth. This option introduces a second equation for the C_f compensation capacitor and an additional test to determine which equation applies. This section describes the gain-peaking effect and the counteracting phase compensation. However, the selection of the final compensation, if any, depends upon analytical tests of the amplifier capacitance effects. The following two sections develop the tests and the selection criteria.

First, consider the gain peaking and the C_f design equation that counteracts it. The peaking results from a new effect introduced by the amplifier input capacitance in the noninverting case. To evaluate this peaking, an examination of the noninverting circuit in Fig. 4.9 reveals the overall input capacitance effects. First, superposition grounding of e_i, to determine ß, still places $C_i = C_{id} + C_{icm}$ in parallel with R_1. Thus the amplifier's input capacitance alters the feedback in the same way as described previously for the inverting case, and the feedback factor ß remains the same. However, the upper C_{icm} component now also affects the circuit's closed-loop gain. Grounding R_1 to form the noninverting configuration, places this resistor permanently in parallel with the upper C_{icm} component. At higher frequencies this C_{icm} shunts R_1, introducing a response zero in the closed-loop gain.

Previously, the inverting configuration avoided this zero by isolating the upper C_{icm} component from the input signal. In Fig. 4.7, feedback holds the amplifier's inverting input at a virtual ground,

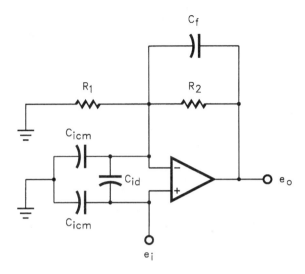

For $f_z < \beta_0 f_c$, add the greater of

$C_f = (R_1/R_2)C_{icm}$, or

$$C_f = (C_c/2)\left[1 + \sqrt{1 + 4C_i/C_c}\right]$$

where $C_c = 1/2\pi R_2 f_c$

and $C_i = C_{id} + C_{icm}$

Figure 4.9 Noninverting configurations connect R_1 directly in parallel with the upper C_{icm} component and impress e_i upon this capacitance.

removing the signal from this capacitance. However, in the noninverting case of Fig. 4.9, feedback reproduces the input signal e_i at the amplifier's inverting input and across the upper C_{icm} capacitance. In this case, that capacitance produces a new component of feedback current. At higher frequencies that current increases the circuit's high-frequency gain. The resulting gain peaking degrades the response accuracy when this effect occurs within the response range of the amplifier. Then the associated peaking often raises the circuit's gain response above the +3-dB limit of a ±3-dB bandwidth criterion.

Fortunately, the phase compensation capacitor C_f also counteracts the gain peaking. This capacitor bypasses R_2, producing a closed-loop response pole, and an appropriate choice of C_f makes this pole cancel the zero produced by C_{icm}. Cancellation results when the break frequency of C_f and R_2 equals that of C_{icm} and R_1. The two capacitances transform the closed-loop gain from $1 + R_2/R_1$ to $1 + Z_2/Z_1$, where $Z_1 = R_1(1 + R_1 C_{icm} s)$ and $Z_2 = R_2(1 + R_2 C_f s)$. Setting the Z_2 break frequency at that of Z_1 makes the Z_2/Z_1 ratio constant with frequency for a flat gain response. Equating the two break frequencies defines the value for C_f as

$$C_f = (R_1/R_2)C_{icm}$$

for compensation of the noninverting gain peaking. As will be described, this design equation replaces the previous C_f expression when the new equation indicates a greater value for C_f.

In practice, the accuracy of this pole-zero cancellation limits the response correction of the gain-peaking compensation. Capacitance variations due to production variances and a voltage sensitivity affect this cancellation. However, the capacitance variations only affect the degree of response correction, and the gain-peaking compensation still improves the response when compared with the uncompensated state. Obviously, production variances in C_{icm} and C_f restrict the predictability of their associated break frequencies. These tolerance variations produce a high-frequency gain shift, higher or lower, in proportion to the net tolerance mismatch.

Less obviously, the voltage dependence of C_{icm} makes the cancellation accuracy a function of the circuit's input signal. The common-mode input capacitance C_{icm} results from transistor junction capacitances that vary with the square root of the applied junction voltages. Input signal e_i exercises these voltages, making C_{icm} a moving target for cancellation. As a result, the circuit's closed-loop gain varies slightly with the signal level at higher frequencies. While generally small, this gain variation with the signal produces high-frequency distortion. This distortion exists with or without the gain-peaking compensation, and the compensation still retains an improvement for bandwidth. Later the compensation of the differential input case described in Sec. 4.2.9 identifies an alternative that cancels the gain peaking of the noninverting case more accurately, and this alternative is developed in Sec. 4.2.10. There the addition of an input resistor produces a gain roll off with one C_{icm} component that cancels the gain peaking produced by the other.

4.2.5 Bandwidth improvement in the noninverting case

Graphical analysis illustrates the gain peaking, defines the bandwidth, and develops a simple test for choosing between the two C_f

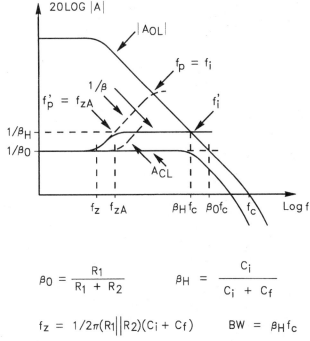

$$\beta_0 = \frac{R_1}{R_1 + R_2} \qquad \beta_H = \frac{C_i}{C_i + C_f}$$

$$f_z = 1/2\pi(R_1\|R_2)(C_i + C_f) \qquad BW = \beta_H f_c$$

Figure 4.10 Modified phase compensation of Fig. 4.9 maintains stability, removes A_{CL} peaking, and extends bandwidth for noninverting cases.

equations developed in the preceding sections. Figure 4.10 supports this analysis for the case where the gain-peaking compensation prevails in the choice of equation. As will be described, this phase compensation often improves the noninverting circuit bandwidth over that of the previous compensation method.

In the figure, the noninverting circuit's initial and final A_{CL} responses appear along with the relevant A_{OL} and 1/ß curves. There the dashed and the solid curves represent A_{CL} and 1/ß for two different phase compensation cases. Consider the dashed-line case first, which represents the initial A_{CL} and 1/ß produced by the phase compensation described for the inverting configuration. For the noninverting configuration considered here, this same compensation assures stability but fails to control the gain peaking. Using that phase compensation in this case results in the dashed A_{CL} response, which rises at higher frequencies. This gain peaking terminates the circuit's response accuracy early in a +3-dB bandwidth limit. The solid curves represent A_{CL} and 1/ß for the final compensation, where an increased C_f removes this peaking. Then the solid A_{CL} response curve remains flat, moving the bandwidth limit out to the f_i' intercept. There the intercept of the solid 1/ß curve with the A_{OL} response defines the new circuit bandwidth. For the case illustrated, the f_i' bandwidth exceeds the f_i bandwidth achieved previously by the inverting configuration.

The improved bandwidth actually results from the C_{icm} shunting of R_1 in the noninverting case. Note that this condition only occurs with high feedback resistances, where this shunting becomes significant within the response range of the op amp. In those cases, the shunting produces both the gain peaking and the bandwidth extension of the example illustrated. Previously, the inverting circuit faced bandwidth limits imposed both by the 1/ß intercept and by the C_f phase compensation. There the C_f shunting of R_2 converted the circuit gain from $-R_2/R_1$ to $-Z_2/R_1$, rolling off the gain with Z_2. Optimum bandwidth resulted from setting this roll off coincident with the f_i intercept. In the noninverting case here, C_f still shunts R_2, but the C_{icm} shunting of R_1 counteracts the C_f effect. Together, the two capacitances convert the noninverting circuit's gain from $1 + R_2/R_1$ to $1 + Z_2/Z_1$. Now the roll off of Z_1 counteracts that of Z_2, removing the Z_2 bandwidth limit. Then the circuit's second bandwidth limit imposed by the intercept prevails and for Fig. 4.10, BW = f_i'.

To quantify the net bandwidth result, note that the gain-peaking compensation levels off the solid 1/ß curve of the figure prior to its intercept with A_{OL}. The resulting level curve simplifies the determination of f_i' in terms of known circuit parameters. Previously, the inverting circuit discussion showed that a straight-line, level extension of the 1/ß$_0$ curve intersects A_{OL} at a frequency of ß$_0 f_c$. Analogously, the solid 1/ß curve here assumes a straight, level value of 1/ß$_H$, making $f_i' = ß_H f_c$, where the quantity ß$_H$ simply represents the high-frequency value of ß. As noted earlier, ß remains the same for the noninverting circuit as for the inverting circuit, and the previous results express the circuit's ß through an equation for 1/ß. Inverting this equation produces

$$ß = \frac{1 + R_2 C_f s}{1 + (R_1 || R_2)(C_i + C_f)s} ß_0$$

where ß$_0 = R_1/(R_1 + R_2)$. At high frequencies, this reduces to ß$_H = C_i/(C_i + C_f)$, making the bandwidth for the case in Fig. 4.10

$$BW = ß_H f_c = \frac{C_i f_c}{C_i + C_f}$$

where $C_i = C_{id} + C_{icm}$. This bandwidth applies for the gain-peaking compensated noninverting case.

For many noninverting amplifiers the gain-peaking compensation does not extend the bandwidth. There the simple stability compensation described with the inverting circuit applies. This sets the bandwidth limit at the same point as described for the inverting case, BW = $1.4f_i$, where $f_i = \sqrt{f_c/2\pi R_2(C_i + C_f)}$ and $C_i = C_{id} + C_{icm}$. The bandwidth equation that applies in a given case depends upon the compensation selection described next.

4.2.6 Selecting the noninverting compensation

A second mathematical test determines which phase compensation method best suits a given noninverting circuit. This test centers around the response zero potentially introduced in the circuit's A_{CL} response by the amplifier's input capacitance at f_{zA}. Referring to Fig. 4.10, the f_{zA} zero in the dashed A_{CL} response only affects the circuit when f_{zA} occurs within the circuit's response range. Beyond the A_{OL} response boundary of the op amp, the circuit's inherent roll off already limits the bandwidth, preventing the gain peaking associated with f_{zA}. As shown in the figure, the dashed-line extension of $1/\beta_0$ defines the response boundary limit of $\beta_0 f_c$ for an amplifier having a low-frequency feedback factor of β_0. Thus gain peaking can only occur for $f_{zA} < \beta_0 f_c$. To define f_{zA}, analysis expands the previous closed-loop gain expression from $1 + Z_2/Z_1$ to

$$A_{CL} = \frac{R_1 + R_2}{R_1} \frac{1 + (R_1 || R_2)(C_{icm} + C_f)s}{1 + R_2 C_f s}$$

This result defines the A_{CL} zero as $f_{zA} = 1/2\pi(R_1 || R_2)(C_{icm} + C_f)$.

Comparison of this zero's frequency with that of the $1/\beta$ response differentiates the two possible phase compensation requirements for the noninverting amplifier. As described previously, the zero in the $1/\beta$ response occurs at $f_z = 1/2\pi(R_1 || R_2)(C_i + C_f)$, where $C_i = C_{id} + C_{icm}$. Then both f_z and f_{zA} depend upon the same resistance combination $R_1 || R_2$. However, the capacitance $C_{id} + C_{icm} + C_f$ sets f_z whereas only $C_{icm} + C_f$ sets f_{zA}, and this capacitance difference ensures that $f_z < f_{zA}$. Thus it often occurs that $f_z < \beta_0 f_c$, requiring phase compensation, but $f_{zA} > \beta_0 f_c$, preventing gain peaking. In those cases, the simple stability correction described for the inverting case also applies for the noninverting amplifier. In other cases, both $f_z < \beta_0 f_c$ and $f_{zA} < \beta_0 f_c$, and the gain-peaking compensation option of the noninverting configuration may benefit the bandwidth.

Further examination of Fig. 4.10 clarifies the phase compensation choice and removes the need to remember the f_{zA} equation. For simple stability compensation, the addition of C_f produces a $1/\beta$ pole at $f_p =$

$1/2\pi R_2 C_f$. Circuit stability requires that this pole be at no higher frequency than f_i. For the example of Fig. 4.10, gain-peaking compensation increases C_f, moving the pole from f_p back to f_p'. There the pole cancels the effect of f_{zA}, removing the gain peaking. However, if $f_{zA} > f_p$, this cancellation requires decreasing C_f, moving the pole forward to an f_p' beyond f_p. Such a move places the pole beyond the maximum frequency f_i set by stability requirements. Thus gain-peaking compensation applies only where $f_{zA} < f_p$, and the alternate compensation requires increasing C_f. This restriction defines the second test in determining the phase compensation requirement imposed by the amplifier input capacitance. Gain-peaking compensation applies only where it specifies a greater C_f value than does the simple stability compensation.

In summary, the determination of a required phase compensation for noninverting amplifiers consists of four steps. First, examine the uncompensated state with $C_f = 0$ and compute $f_z = 1/2\pi(R_1||R_2)C_i$, where $C_i = C_{id} + C_{icm}$. Compare the resulting f_z with $\beta_0 f_c$, where $\beta_0 = R_1/(R_1 + R_2)$. The condition $f_z > \beta_0 f_c$ indicates stability and no gain peaking without phase compensation. The condition $f_z < \beta_0 f_c$ indicates the need for phase compensation. If the latter occurs, first compute the C_f values prescribed by both compensation methods. For simple stability compensation,

$$C_f = \frac{C_c}{2}\left(1 + \sqrt{1 + \frac{4C_i}{C_c}}\right)$$

where $C_c = 1/2\pi R_2 f_c$ and $C_i = C_{id} + C_{icm}$. For gain-peaking compensation,

$$C_f = (R_1/R_2)C_{icm}$$

Then compare the two C_f values found and use the larger of the two. When the gain-peaking compensation specifies a larger value, the added capacitance improves the bandwidth.

4.2.7 Input capacitance and the differential input configuration

A new component of op amp input capacitance affects the differential input configuration or difference amplifier. This configuration also suggests conflicting phase compensation demands due to its combined inverting and noninverting gain paths. Fortuitously, a logical choice of circuit resistances turns the added capacitance problem into the solution for the compensation conflict. This coincidence reduces the phase compensation evaluation and selection to that described earlier for the inverting configuration. However, the differential input configuration requires two matching compensation capacitors to retain the impedance balance required for good common-mode rejection.

Figure 4.11 shows the generalized difference amplifier with the op amp input capacitances separated for analysis. There the addition of a second C_f capacitor retains the circuit balance fundamental to the configuration. At first this circuit suggests a complex phase compensation task. The circuit presents both inverting and noninverting gain paths, suggesting conflicting demands between simple stability compensation and gain-peaking compensation. Also, this circuit makes the lower C_{icm} component a factor in the circuit's response. The differ-

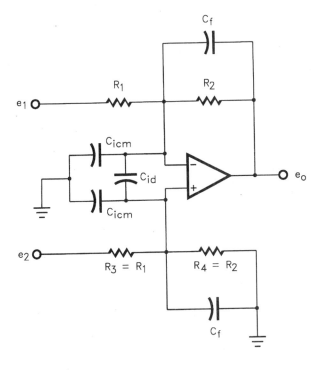

For $f_z < \beta_0 f_c$, add

$$C_f = (C_c/2)\left[1 + \sqrt{1 + 4C_i/C_c}\right]$$

where $C_c = 1/2\pi R_2 f_c$

and $C_i = C_{id} + C_{icm}$

Figure 4.11 For difference amplifiers, balanced resistances and capacitances produce cancellation of major input capacitance effects.

ence amplifier connection places R_4 directly in parallel with this capacitance, resulting in a new roll off in the e_2 signal path. No equivalent roll off affects the e_1 signal path, and different high-frequency gains for the two signals warn of degraded common-mode rejection.

However, the balanced nature of op amp inputs and the logical choice of resistors combine to resolve all of these issues. Then the new roll off of the e_2 gain path accurately compensates for the gain peaking described for the noninverting case. As a result, the gains remain matched in the two signal paths, preserving common-mode rejection. This circuit's inherent compensation for gain peaking also resolves the conflicting demands of the two phase compensation methods described before. Phase compensation for the well-designed difference amplifier need only assure stability, with no special consideration for gain peaking. Analysis of the circuit's closed-loop response demonstrates these fortuitous conditions.

For simplicity, the corresponding A_{CL} analysis neglects the roll-off effects of the amplifier's A_{OL} response. That roll off adds an A_{CL} pole at the f_i intercept, independent of the response compensation described here. Neglecting that pole, first assume the uncompensated state, with $C_f = 0$. Then analysis of the circuit shows that

$$A_{CL} = -\frac{R_2}{R_1} e_1 + \frac{R_4(R_1 + R_2)}{R_1(R_3 + R_4)} \frac{1 + (R_1||R_2)C_{icm}s}{1 + (R_3||R_4)C_{icm}s} e_2$$

This generalized difference amplifier expression suggests greatly different gains, both dc and ac, for the e_1 and e_2 signals. Yet the primary benefit of the difference amplifier, common-mode rejection, requires balanced gain magnitudes for these two signals. When the gain magnitudes match, the opposite polarities of the two gains cancel any signal common to e_1 and e_2.

4.2.8 Balancing the differential input configuration

Resistance selection develops the gain balance described, first through the resistance ratio and then through the resistance value. Consider first the uncompensated case with the C_f capacitors set to zero. For gain balance, the difference amplifier imposes a fundamental resistor ratio requirement of $R_4/R_3 = R_2/R_1$, reducing the closed-loop response to

$$A_{CL} = -\frac{R_2}{R_1} e_1 + \frac{R_2}{R_1} \frac{1 + (R_1||R_2)C_{icm}s}{1 + (R_3||R_4)C_{icm}s} e_2$$

Setting s to zero in this equation produces the dc result with the same R_2/R_1 gain magnitude for the two signal gains.

However, the ac gains still differ due to the pole and zero in the e_2 response. These singularities develop at $f_{p2} = 1/2\pi(R_3||R_4)C_{icm}$ and $f_{z2} = 1/2\pi(R_1||R_2)C_{icm}$. Ac continuation of balanced circuit gains requires matching these two break frequencies. Then a pole-zero cancellation removes the effects of f_{p2} and f_{z2}. As written, both singularities depend upon the capacitance C_{icm}, but the two actually result from different capacitances of the op amp. In the figure, the lower C_{icm} component produces f_{p2} and the upper C_{icm} component f_{z2}. Fortunately, the balanced nature of op amp inputs matches these capacitances very closely, and they are assumed to be equal. The two singularities also depend upon the resistances $R_3||R_4$ and $R_1||R_2$. Matching these resistances, along with the earlier $R_4/R_3 = R_2/R_1$ requirement, simply requires that $R_3 = R_1$ and $R_4 = R_2$. Under these conditions, the difference amplifier's added e_2 roll off cancels with the circuit's gain peaking for

$$A_{\text{CL}} = -\frac{R_2}{R_1}e_1 + \frac{R_2}{R_1}e_2$$

The resulting equal gain magnitudes, combined with the opposite gain polarities, assure good common-mode rejection.

4.2.9 Compensating the differential input configuration

Phase compensation disturbs the impedance balance achieved by this resistance matching. However, the logical continuation of impedance balancing restores common-mode rejection. When required, adding the C_f bypass of R_2 restores stability as described previously with the inverting configuration. This bypass adds the same pole in both the e_1 and the e_2 gain paths at $1/2\pi R_2 C_f s$, leaving gain matching undisturbed. However, this C_f also alters f_{z2} of the e_2 response with no matching effect upon the e_1 response. This f_{z2} zero develops with the net impedance presented to the upper C_{icm} component. Without phase compensation, the feedback network presents the impedance $R_1||R_2$ to this capacitance. Adding the C_f bypass of R_2 changes this to $R_1||R_2||(1/C_f s)$, moving f_{z2} to $1/2\pi(R_1||R_2)(C_{icm} + C_f)$. There it no longer cancels with the pole still at $f_{p2} = 1/2\pi(R_1||R_2)C_{icm}$.

This pole must be adjusted to realign the previous pole-zero cancellation. Examination of these last f_{z2} and f_{p2} expressions suggests an additional capacitance C_f for this adjustment, and this addition supports the obvious correction for the difference amplifier circuit where impedance balance governs this circuit's common-mode response. Any bypass of R_2 must be matched by a balancing bypass of R_4, as provided by the lower C_f capacitor of Fig. 4.11. This second compensation capacitor moves f_{z2} to $f_{p2} = 1/2\pi(R_1||R_2)(C_{icm} + C_f)$, restoring the pole-zero cancellation. Then for the difference amplifier with matched resistances and capacitances,

$$A_{CL} = \frac{R_2}{R_1(1 + R_2 C_f s)} (e_2 - e_1)$$

Under these balanced conditions, the difference amplifier automatically avoids the gain peaking described for the noninverting configuration, reducing the phase compensation selection to its most basic case. The balanced impedances assure that the gain roll off of the e_2 signal path accurately counteracts the gain peaking described for the noninverting configuration. This removes the previously mentioned conflict between the combined inverting and noninverting compensation requirements of the difference amplifier. For the balanced impedances described, phase compensation selection for the effects of amplifier input capacitance reverts to the simple stability restoration of the inverting case.

As described for that configuration, stability restoring phase compensation first involves a test for the requirement and then a calculation of any required compensation. An affirmative result in the $f_z < \beta_0 f_c$ test indicates the need for compensation. In this test, assume the uncompensated state with $C_f = 0$ and calculate $f_z = 1/2\pi(R_1||R_2)C_i$, where $C_i = C_{id} + C_{icm}$. Also in this test, note that $\beta_0 = R_1/(R_1 + R_2)$, as described previously. For affirmative test results, add the two C_f compensation capacitors shown with capacitance values defined by the C_f equation of the inverting circuit discussion. From that earlier discussion,

$$C_f = \frac{C_c}{2} \left(1 + \sqrt{1 + \frac{4C_i}{C_c}}\right)$$

where $C_c = 1/2\pi R_2 f_c$ and $C_i = C_{id} + C_{icm}$.

This phase compensation affects both gain paths in the same manner, producing equal bandwidths for the two. As described for the inverting configuration, choosing C_f with the preceding equation places the bandwidth-limiting pole in coincidence with a similar feedback limit. Previously, Fig. 4.8 illustrated this coincidence where the pole at f_p coincides with the $1/\beta$ and A_{OL} intercept at f_i. The response curves of the earlier figure apply to the difference amplifier as well. The coincidence of the two bandwidth limits produces slight gain peaking, moving the -3-dB bandwidth limit to BW $= 1.4 f_i$. As developed earlier, $f_i = \sqrt{f_c/2\pi R_2(C_i + C_f)}$ and $C_i = C_{id} + C_{icm}$.

4.2.10 Alternative noninverting compensation

The pole-zero cancellation described for the difference amplifier also presents an alternative for the noninverting configuration. In the simplest case, grounding the e_1 input of the difference amplifier converts the circuit to a noninverting amplifier. However, this simple approach fails to provide the high input resistance expected of a noninverting circuit. In that condition, the difference amplifier of Fig. 4.11 presents an input resistance of just $R_3 + R_4$ at the noninverting e_2 input.

Examination of the difference amplifier response equations identifies a noninverting alternative that produces the desired input resistance. This alternative also solves an input capacitance problem and a distortion problem not resolved by the earlier noninverting compensation. With a balanced difference amplifier, pole-zero cancellation automatically aligns the response singularities f_{p2} and f_{z2} of the noninverting signal path described before. After adding phase compensation, these singularities become $f_{p2} = 1/2\pi(R_3 || R_4)(C_{icm} + C_f)$ and $f_{z2} = 1/2\pi(R_1 || R_2)(C_{icm} + C_f)$. Simply making $R_3 || R_4 = R_1 || R_2$ aligns this pole and zero for the difference amplifier. However, the difference amplifier imposes a second resistance requirement, $R_2/R_1 = R_4/R_3$, to match the circuit's two gain magnitudes. Combined, the two resistance requirements force the choices $R_3 = R_1$ and $R_4 = R_3$, resulting in low input resistance at the noninverting input.

The single gain of the noninverting amplifier removes the gain-matching requirement, reducing the resistance-matching requirement to $R_3 || R_4 = R_1 || R_2$ for pole-zero alignment only. In the simplest case, $R_4 = \infty$ and $R_3 = R_1 || R_2$, restoring the $R_3 + R_4$ input resistance to a high level. Figure 4.12 shows the result with a modification required to accommodate higher source resistance. There reducing R_3 to $R_1 || R_2 - R_S$ adjusts for the presence of R_S, keeping the net resistance equal to $R_1 || R_2$. This resistance balance again assures the pole-zero cancellation required for response flatness.

Including the effect of R_S in this balance also counteracts the reaction of source resistance with amplifier input capacitance. The previous noninverting compensation excludes R_3 and the lower C_f, ignoring the source resistance effect. Without these components, the lower C_{icm} capacitance and R_S of Fig. 4.12 produce an uncompensated response pole in the e_2 gain path. This pole does not align with the response zero produced by the upper C_{icm}. Together, the pole and zero potentially produce higher-frequency deviations in the circuit's closed-loop response. Higher source resistances produce these deviations, making the alternative compensation, here, the preferred choice. Then adding R_3 and C_f realigns the pole and zero, restoring response flatness. However, this alternative requires two more circuit components and does not realize the full bandwidth advantage of the previous gain-peaking compensation. Thus the gain-peaking compensation remains the preferred choice for lower source resistance cases.

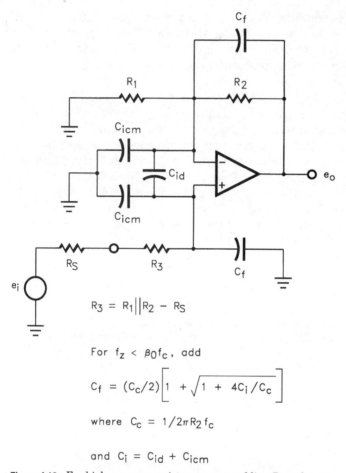

Figure 4.12 For higher source resistance cases, adding R_3 and a second C_f to noninverting configurations produces more exact cancellation of C_{icm} effects.

$$R_3 = R_1 \| R_2 - R_S$$

For $f_z < \beta_0 f_c$, add

$$C_f = (C_c/2)\left[1 + \sqrt{1 + 4C_i/C_c}\right]$$

where $C_c = 1/2\pi R_2 f_c$

and $C_i = C_{id} + C_{icm}$

The pole-zero cancellation of the alternative compensation also removes a distortion effect described previously. The voltage sensitivity of the upper C_{icm} produces distortion with the earlier noninverting circuit. Signal voltage on this junction capacitance varies the capacitance with the signal. This varies the response zero produced by the capacitance, altering the high-frequency gain as the signal magnitude varies. With the alternative compensation here, the pole-zero cancellation tracks the zero's variation. Both the zero and the pole result from C_{icm} components, and the two components inherently match closely. Further, equal signal voltages on the two capacitances assure a capacitance match undisturbed by signal variation. The signal at the amplifier's noninverting input drives the lower C_{icm} component. Feedback replicates this signal at the amplifier's inverting input, producing a matching signal on the upper C_{icm} component. Both capacitances vary with the impressed signal, but by the same amount. Thus variations in the pole's frequency track those of the zero's, maintaining the pole-zero cancellation. Removing the high-frequency signal dependence of the circuit's gain avoids the associated distortion described for the previous noninverting circuit.

The resistance balance of this compensation method also cancels the dc offset effect of the op amp's input bias currents. The amplifier's two equal input currents now flow into equal net resistances $R_1||R_2$ and $R_3 + R_S = R_1||R_2$, and the resulting offset effects cancel in the circuit's output response.

The phase compensation selection and the bandwidth calculation for the noninverting alternative here duplicate those described for the difference amplifier. First, the $f_z < ß_0 f_c$ test determines whether or not the circuit requires phase compensation. For this test, $ß_0 = R_1/(R_1 + R_2)$ and the uncompensated test case makes $C_f = 0$ for $f_z = 1/2\pi(R_1||R_2)(C_{id} + C_{icm})$. When $f_z < ß_0 f_c$, add the two C_f compensation capacitors shown with values prescribed by

$$C_f = \frac{C_c}{2}\left(1 + \sqrt{1 + \frac{4C_i}{C_c}}\right)$$

where $C_c = 1/2\pi R_2 f_c$ and $C_i = C_{id} + C_{icm}$. This phase compensation follows directly from the earlier inverting circuit discussion and focuses upon stability without regard for gain peaking. As with the difference amplifier, this noninverting alternative automatically removes the gain peaking. As before, this compensation produces a closed-loop bandwidth of BW = $1.4f_i$, where $f_i = \sqrt{f_c/2\pi R_2(C_i + C_f)}$ and $C_i = C_{id} + C_{icm}$.

4.3 Multipurpose Phase Compensation

The previous two sections describe phase compensation for the secondary poles created by two specific capacitance effects. Other phase compensation techniques offer correction for any source of secondary poles. These more general techniques build upon the feedback-factor compensation introduced in the previous section and again tailor the circuit's feedback factor, rather than the A_{OL} response. While more traditional phase compensation modifies the A_{OL} response, most op amps lack a provision for such modification. The feedback modifications described here produce equivalent stability results utilizing external circuit options.

Fundamentally, the phase compensation techniques of this section shape the 1/ß curve to move its intercept with the A_{OL} response to a stable region. They do so by intentionally introducing a 1/ß rise at higher frequencies. In the last section, the amplifier input capacitance produced just such a rise, which jeopardized the circuit's rate of closure at the intercept. However, such a rise also moves the intercept up and away from secondary A_{OL} poles. The feedback-factor phase compensation takes advantage of this repositioning to accommodate any source of secondary amplifier poles. Examples demonstrate this technique for both negative and positive feedback as well as the special case of the integrator circuit.

These multipurpose compensation techniques also access higher slew rates for low-gain applications. There the feedback-factor compensation permits the use of lightly compensated op amps intended for higher gains. Numerous op amp product models offer the choice of two internal compensation options. With one option, unity-gain compensation assures frequency stability for any gain level but restricts the slew rate. With the other, lighter phase compensation reduces the slew-rate restriction but imposes a minimum gain for stable operation. These internal compensation alternatives tailor the open-loop response to optimize the compromise of minimum gain versus slew rate of conventional phase compensation. In this process, adjusting the open-loop response to permit stable low-gain operation simultaneously reduces the slew rate. Feedback-factor compensation tailors 1/ß instead, leaving the slew rate unaffected. Thus an op amp with lighter internal compensation, supplemented by external feedback-factor compensation, extends the higher-slew-rate option to lower gains as well.

4.3.1 Feedback-factor compensation with negative feedback

Figure 4.13 adds feedback-factor phase compensation to the negative feedback case through the added R_C and C_C. Shown there with a noninverting amplifier, this method also applies to all other op amp configurations. In the special case of the voltage follower, the technique requires the addition of resistor R_2 as part of the compensation. This phase compensation addresses a two-pole amplifier response of any origin. The second pole can be the result of capacitive loading, parasitic input capacitance, or a lightly phase-compensated amplifier. Previously described compensation methods lack this versatility because they only address specific sources of secondary poles.

The phase compensation elements of this method connect between the op amp inputs to alter the circuit's feedback factor without altering the ideal closed-loop gain. Elements R_C and C_C are bootstrapped

Phase Compensation   141

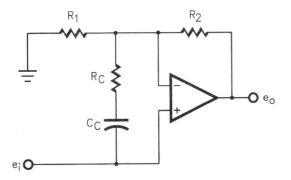

$$C_C = 5/\pi R_C f_{p2} \qquad R_C = \frac{R_2}{A_{min} - 1/\beta_0} \qquad \beta_0 = \frac{R_1}{R_1 + R_2}$$

Figure 4.13 A series R-C network between op amp inputs adds a phase compensation pole and zero to circuit feedback factor.

on the input signal e_i, and they produce no feedback current in direct response to e_i. Thus these compensation elements produce no current that would alter the circuit's gain. Further, the circuit still impresses the signal e_i upon the R_1 feedback element, since feedback forces the two op amp inputs to the same voltage. There e_i creates a feedback current that produces a corresponding voltage on R_2 and an output signal e_o. Thus R_1 and R_2 continue to control the ideal closed-loop gain applied to input signal e_i.

However, the circuit does impress the feedback error signal, developed between the op amp inputs, upon the R_C and C_C combination. There, and in R_1, the error signal produces residual feedback currents that control bandwidth and stability. All three elements contribute to the gain applied to the feedback error signal, and thus all three contribute to the circuit's net feedback factor. These three elements form a voltage divider with R_2, and the resulting divider ratio equals the feedback factor. Then $\beta = (R_1 || Z_C)/(R_1 || Z_C + R_2)$, where $Z_C = R_C + 1/C_C s$. At low frequencies, Z_C remains very large and β reduces to the familiar $\beta_0 = R_1/(R_1 + R_2)$. At higher frequencies, the addition of Z_C introduces a frequency dependence to β, tailoring the $1/\beta$ response. Figure 4.14 shows the result, where a zero followed by a pole raises $1/\beta$ at higher frequencies and then levels it. First C_C introduces the response zero to produce a rising $1/\beta$ curve. Then R_C introduces the response pole to level out the $1/\beta$ curve prior to the f_i intercept. This raises the $1/\beta$ intercept to a region of stable intersection with A_{OL}. Without C_C and R_C, the $1/\beta_0$ curve shown would continue on to an intercept in a region of two-pole gain slope, as shown by a dashed line.

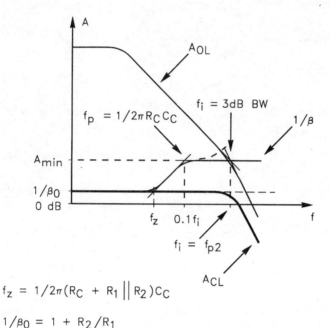

$f_z = 1/2\pi(R_C + R_1 || R_2)C_C$

$1/\beta_0 = 1 + R_2/R_1$

Figure 4.14 The pole and zero added to Fig. 4.13 raise the 1/ß curve to a region of stable intercept with A_{OL}.

Intuitive analysis defines the locations of the 1/ß zero and pole. The zero, f_z, results when C_C breaks with the net resistance presented to this capacitance. This resistance includes R_C plus the resistance presented at the R_1 and R_2 junction. The latter two resistors both return to low impedances, so they are effectively in parallel. Then

$$f_z = 1/2\pi(R_C + R_1 || R_2)C_C$$

Following this response zero, the impedance of C_C continues to decline, causing 1/ß to rise. This impedance decline terminates upon reaching the minimum impedance set for Z_C by R_C. This produces a 1/ß response pole at $f_p = 1/2\pi R_C C_C$ and levels off the 1/ß curve.

4.3.2 Side effects of feedback-factor compensation

Feedback-factor compensation with negative feedback introduces two limitations. This compensation remains vulnerable to amplifier input capacitance, and the compensation elements reduce the circuit's input impedance. Amplifier input capacitance produces a second 1/ß rise and sometimes compromises frequency stability. Shown as a rising dashed curve in Fig. 4.14, this 1/ß rise results from the bypass of the

R_C and C_C phase compensation. There the op amp's input capacitance shunts this compensation network, overriding the 1/ß leveling effect of R_C. A more detailed discussion of op amp input capacitance effects appears in Sec. 4.2. This second 1/ß rise compromises the circuit's rate of closure at the critical intercept if the rise develops at a low enough frequency. That occurs when the break frequency of $R_C + R_1||R_2$ and $C_i = C_{id} + C_{icm}$ precedes the intercept at f_i. Here C_{id} and C_{icm} represent the differential and common-mode input capacitances of the amplifier. Lower feedback resistance levels move this break frequency beyond f_i, avoiding the problem.

The phase compensation in Fig. 4.13 also reduces the input impedance for noninverting applications. At first, a very low impedance might be expected because the compensation elements connect directly to the circuit input. However, as described earlier, the circuit bootstraps these elements on the input signal. No input current flows from them in direct response to e_i. Indirectly, e_i creates a signal on the R_C and C_C combination through the gain error signal between the op amp inputs. That signal results from the finite open-loop gain of the op amp and equals $e_o/A_{OL} = A_{CL}e_i/A_{OL}$. Here A_{OL} and A_{CL} are the open-loop and closed-loop gains of the circuit. This error signal produces a current $A_{CL}e_i/A_{OL}Z_C$ in $Z_C = R_C + 1/C_Cs$, and the Z_C connection delivers this current to the circuit input, defining an input impedance of

$$Z_I = A_{OL}Z_C/A_{CL}$$

where $Z_C = R_C + 1/C_Cs$. Thus the bootstrap of Z_C boosts the resulting input impedance by the loop gain A_{OL}/A_{CL}.

Shaping the 1/ß response phase compensates circuits for parasitic poles or adapts lighter-compensated op amps to lower-gain applications. For the latter applications, the lighter-compensated amplifiers deliver a greater slew rate, but they do not increase the bandwidth with this compensation technique. At first, greater circuit bandwidth might also be expected because the lighter amplifier compensation produces greater open-loop bandwidth. However, such amplifiers impose a minimum stable gain for the 1/ß intercept. Raising the 1/ß curve to accommodate this requirement relinquishes the greater bandwidth. As a result, the circuit bandwidth typically remains the same as with a unity-gain compensated version of the op amp.

The higher slew rate achieved in that case tends to improve the settling time, but the basic feedback-factor compensation described negates this improvement. There the addition of f_z and f_p to the 1/ß response of Fig. 4.14 introduces a long settling tail. Alternately, eliminating C_C removes the added pole and zero to avoid this settling time limitation, and R_C alone raises the 1/ß curve for a stable intercept. However, this alternative raises the 1/ß response at all frequencies, increasing the gain supplied to the amplifier's input offset voltage and lower-frequency noise.

4.3.3 Selecting the negative feedback components

Component value selection for the phase compensation in Fig. 4.13 focuses upon the second A_{OL} pole. Whatever the cause of this pole, its position defines a minimum amplifier gain A_{min} for which the amplifier displays good stability. Note that for this application, A_{min} reflects the minimum level of A_{OL} for the 1/ß intercept and not the minimum A_{CL} for the circuit. Placing the 1/ß intercept at that level usually places the intercept at the second pole, and this assures around 45° of phase margin. Placing the intercept prior to the second pole would increase the phase margin, but this placement would require raising the 1/ß curve further at high frequencies. That would also increase the 1/ß gain received by the amplifier's high-frequency noise and move f_i to a lower frequency, reducing the bandwidth.

In compromise, the selection of R_C places the critical intercept at the frequency of the second A_{OL} pole. To do so, the choice of R_C sets the high-frequency 1/ß level equal to A_{min}. In the case of a lightly compensated amplifier, A_{min} is a specified value, and for other cases, measurement or calculation defines this value. Whatever its cause, just knowing the location of the second A_{OL} response pole f_{p2} defines A_{min} for a given application. Then the frequency f_{p2} yields A_{min} through the approximation $A_{OL} = f_c/f$. Setting $A_{OL} = A_{min}$ at $f = f_{p2}$ produces $A_{min} = f_c/f_{p2}$ and defines a minimum level for the 1/ß intercept. Similarly, for capacitance loading cases, accurate knowledge of the amplifier output resistance R_o yields A_{min} through the expression $f_{p2} = 1/2\pi R_o C_L$. Substituting this f_{p2} expression in the last A_{min} expression produces $A_{min} = 2\pi R_o C_L f_c$. Here the complex nature of the amplifier's actual R_o, described in Sec. 4.1.3, may require empirical tuning of the phase compensation. Whatever the case, setting the high-frequency value of 1/ß equal to A_{min} places the f_i intercept at the second A_{OL} pole. This phase compensation produces a high-frequency 1/ß of $1/ß_0 + R_2/R_C$. Setting this equal to A_{min} and solving for R_C yields

$$R_C = \frac{R_2}{A_{min} - 1/ß_0} = \frac{R_2}{f_c/f_{p2} - 1/ß_0}$$

where $1/ß_0 = 1 + R_2/R_1$.

Knowing R_C, the C_C selection follows. These two components join in setting the pole and zero, f_p and f_z, of the 1/ß curve of Fig. 4.14. An appropriate choice of C_C ensures that f_p and f_z do not disturb the phase conditions at the f_i intercept. For this purpose, making $f_p = 0.1f_i$ sufficiently separates both f_p and f_z from f_i. Then the phase contributions of f_p and f_z are both fully developed and cancel at the frequency f_i. For $f_p = 1/2\pi R_C C_C = 0.1f_i$ the compensation capacitor becomes $C_C = 5/\pi R_C f_i$. Combining this interim result with two others defines C_C in terms of the A_{min} discussed. First, placing the intercept at f_{p2} sets $f_i = f_{p2}$, making $C_C = 5/\pi R_C f_{p2}$. Next, the preceding R_C discussion showed that $A_{min} = f_c/f_{p2}$. Then $f_{p2} = f_c/A_{min}$, and finally

$$C_C = 5/\pi R_C f_{p2} = 5A_{min}/\pi R_C f_c$$

4.3.4 Feedback-factor compensation for the integrator

A special case of the previous phase compensation provides a higher slew rate to the integrator circuit.[6] Normally, the phase compensation requirement of the op amp limits the integrator's output slewing because that configuration requires an amplifier having unity-gain phase compensation. This results because the short-circuit feedback of the integrator's capacitor produces a unity feedback factor at higher frequencies. At first, this would preclude access to the increased slew rate offered by amplifiers having lighter phase compensation. However, a modified feedback-factor compensation removes this limitation for the integrator.

Figure 4.15 shows that phase compensation, which simply eliminates the R_C element of the previous compensation. In this case, just the capacitor C_C provides the required adjustment in the feedback

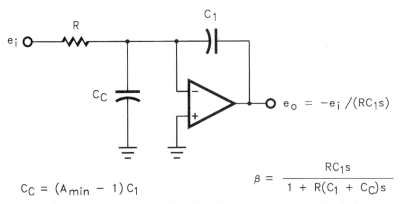

Figure 4.15 Compensation capacitor C_C reduces integrator feedback factor to accommodate a lightly compensated, faster op amp.

factor. Without C_C, the conventional integrator encounters a short-circuit feedback through C_1 at high frequencies, where the impedance of this capacitor becomes far less than that of resistor R. Then the feedback factor $\beta = R/(R + 1/C_1 s)$ rises to unity, requiring a unity-gain compensated op amp. The addition of C_C terminates this rise at a level less than unity to remove this restriction. Then at higher frequencies, the low impedance of C_1 no longer competes with just the impedance of resistor R in the control of the circuit's feedback factor. In the feedback factor analysis of this alternative, superposition grounds the circuit input, placing R and C_C in parallel. Then C_C bypasses the resistor, counteracting the feedback dominance of C_1. The resulting high-frequency feedback factor still rises, but only to the fraction $C_1/(C_1 + C_C)$ rather than to unity. This removes the unity-gain stability requirement for the op amp, and lighter compensated amplifiers, with higher slew rates, satisfy the new circuit's stability requirement.

The addition of C_C leaves the basic integrator response unchanged, but does degrade noise performance. As in the basic integrator case, the input signal e_i still drives resistor R to develop the normal feedback current e_i/R. This current reacts with the impedance of C_1, producing the desired output voltage of $-e_i/(RC_1 s)$. The input signal does not directly drive C_C, so this capacitor does not significantly add to the feedback current. However, the feedback factor reduction produced by C_C increases the noise gain, as illustrated in Fig. 4.16,

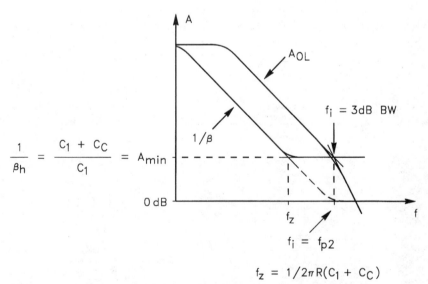

Figure 4.16 Compensation capacitor of Fig. 4.15 levels off integrator $1/\beta$ curve above the unity-gain axis, producing a stable intercept.

through the basic integrator's 1/ß curve and the modification introduced by C_C. Normally, the 1/ß curve would continue its downward slope to the unity-gain axis, as shown by a dashed line. There the curve would intercept A_{OL} in a region of two-pole roll off, producing oscillation in this lightly compensated case. The addition of C_C interrupts the downward 1/ß slope at a level of $(C_1 + C_C)/C_1$, and this moves the 1/ß intercept with A_{OL} upward to a region of reduced A_{OL} slope, restoring stability.

Optimum stability results from setting the intercept f_i well before the second amplifier pole f_{p2}. However, this choice reduces the circuit's bandwidth excessively, as also defined by f_i. Further, reducing f_i raises the 1/ß gain supplied to noise and other op amp errors at higher frequencies. As described in Chap. 1, the associated noise gain follows the 1/ß curve up to the f_i intercept and then rolls off with the A_{OL} response. To preserve bandwidth and minimize this error gain, excessive C_C capacitance should be avoided.

As in previous examples, the optimum choice for C_C aligns the high-frequency 1/ß intercept with the amplifier's minimum stable gain A_{min}. The previous section describes this gain and its direct correspondence to f_{p2}. Placing the 1/ß intercept at f_{p2} aligns the high-frequency $1/ß_H$ with A_{min}, producing 45° of phase margin for good frequency stability without an excessive error gain increase. The high-frequency feedback factor $ß_H$ meets this optimum condition when

$$\frac{1}{ß_H} = \frac{C_1 + C_C}{C_1} = A_{min}$$

Solving for C_C defines the optimum compensation value as

$$C_C = (A_{min} - 1)C_1$$

With this choice for C_C, the circuit's bandwidth remains the same as with the equivalent, conventional integrator. In the conventional case, the heavier phase compensation required of the amplifier reduces the bandwidth by the same amount that C_C does here. Thus the lighter-compensated amplifier considered here delivers the same final bandwidth of f_i that the heavier-compensated version provides at unity gain. However, the lighter-compensated amplifier here provides a higher slew rate to the integrator.

At first it might seem that the addition of C_C would also compromise the frequency stability. This capacitor provides a return to ground for C_1, producing a capacitance load on the op amp output, and this load becomes $C_L = C_1 C_C/(C_1 + C_C)$. However, this load remains small, due to a practical circuit limitation. To benefit from the higher slew rate of this option, the integrator capacitance C_1 must be kept small. Like the op amp's slew-rate limit, this capacitance imposes a limit to the circuit's output rate of change. The circuit drives C_1 with a charging current of e_i/R, causing e_o to change at a maximum rate of $de_o/dt = e_i/RC_1$. Thus a large value for C_1 imposes a rate-of-change limit of its own. In practice, any integrator capable of a de_o/dt approaching the op amp slew-rate limit must have a small value for C_1, and such a value presents little capacitance loading to the op amp.

4.3.5 Positive feedback compensation of voltage followers

For the voltage follower, the previous feedback-factor compensation approach requires the addition of resistance in the negative feedback path. However, such resistance reacts with the op amp's input capacitance, as described in Sec. 4.2.1. This reaction introduces a feedback pole, potentially degrading stability. Alternately, positive feedback avoids this pole and supplies an otherwise equivalent feedback-factor compensation for the voltage follower. Also this positive feedback approach offers the only external phase compensation available for committed voltage followers. For those circuits, internal feedback commits an op amp to the follower role and precludes the addition of resistance in the negative feedback path.

This positive feedback option again tailors the 1/ß curve to phase compensate any source of secondary amplifier poles. For this positive feedback case, note that the circuit's 1/ß curve responds to the net circuit feedback, not just the negative feedback that may be committed internally. Adding positive feedback modifies the net circuit feedback factor to ß = ß_ − ß_+, as described in Sec. 2.2.3. Here ß_ and ß_+ are the individual negative and positive feedback factors. Previously, the feedback-factor phase compensation simply modified ß_, but positive feedback introduces a ß_+ term here to alter the net factor instead.

Figure 4.17 illustrates the positive feedback phase compensation along with the relevant amplifier input capacitance C_{icm}. Compensation elements R_C and C_C again form a feedback voltage divider with resistor R_2. However, here R_2 moves to the input circuit. There R_2 joins R_C and C_C in a positive feedback network around an otherwise normal voltage follower. The relevant amplifier input capacitance C_{icm} again alters the feedback factor, but aids rather than degrades stability. As in the noninverting amplifier case of Fig. 4.13, bootstrap drives the feedback compensation network, limiting any disturbance to the ideal voltage follower performance. As shown, the bootstrap connection again places R_C and C_C in parallel with the op amp's inputs. There only the input error signal of the amplifier appears across the R_C and C_C combination. This small error voltage

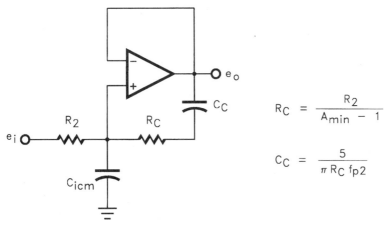

Figure 4.17 With positive feedback phase compensation, the shunting of amplifier input capacitance increases, rather than decreases, net feedback factor.

produces little signal current, minimizing the associated signal voltage developed on R_2. Thus the input signal e_i transfers almost totally to the op amp's noninverting input, and there the signal drives a normal voltage follower for $e_o \approx e_i$ as desired.

Closer examination of the signal across R_C and C_C reveals a bootstrap boost to the input impedance. Low impedance might first be expected due to the direct e_i connection to the feedback network. However, with $e_o \approx e_i$, both ends of the R_C/C_C positive feedback network receive almost the same signal drive. This reduces the current developed in the network dramatically, increasing the circuit's input impedance. Note that this only applies to the voltage follower case with $e_o \approx e_i$, where only the gain error signal of the op amp drops across R_C and C_C. This signal, $e_o/A_{OL} \approx e_i/A_{OL}$, develops between the op amp inputs and across $Z_C = R_C + 1/C_C s$. The resulting current, $e_i/A_{OL}Z_C$, flows through R_2 to the circuit input. Dividing the input signal voltage e_i by this input current defines the circuit's input impedance as $Z_I = A_{OL}Z_C$. Thus the bootstrap drive of the positive feedback network retains high input impedance for this voltage follower option.

The net feedback factor of Fig. 4.17 produces almost the same stability result as the previous negative feedback case. The only difference lies in the effect of the amplifier input capacitance C_{icm}, as described later. Initially neglecting this capacitance, the net factor ß = ß_ − ß_+ here develops from the individual factors of ß_ = 1 and ß_+ = $R_2/(R_2 + Z_C)$, where $Z_C = R_C + 1/C_C s$. This combination produces

$$ß = \frac{1 + R_C C_C s}{1 + (R_C + R_2)C_C s}$$

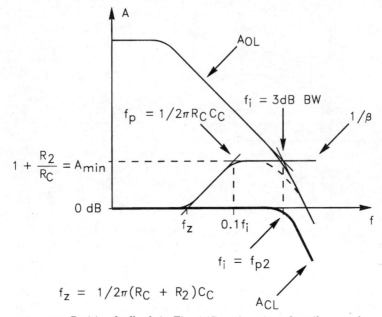

Figure 4.18 Positive feedback in Fig. 4.17 again raises the 1/ß curve, but amplifier input capacitance now produces the 1/ß roll off of the dashed curve.

Figure 4.18 displays the corresponding 1/ß and A_{OL} curves, where the 1/ß curve begins at the unity-gain axis, due to the follower connection of the negative feedback. From there 1/ß experiences a zero and a pole, just as in the previous negative feedback case. The zero occurs at $1/2\pi(R_C + R_2)C_C$ and the pole develops at $1/2\pi R_C C_C$. These response singularities lift the 1/ß curve prior to the f_i intercept, moving that intercept to a region of reduced rate of closure with the A_{OL} curve.

However, including the effect of C_{icm} distinguishes this positive feedback compensation from the earlier negative feedback solution. The positive feedback here responds to this amplifier input capacitance with an opposite change in 1/ß slope. Previously, the negative feedback compensation responded with a zero in the 1/ß response, as shown earlier by a dashed curve in Fig. 4.14. There the resulting zero potentially compromised stability by increasing the circuit's rate of closure at the intercept. For the positive feedback case considered here, the corresponding dashed curve of Fig. 4.18 displays a 1/ß pole rather than a zero, and this pole reduces the rate of closure to improve rather than compromise stability. This difference originates in the net feedback factor ß = ß_ − ß_+, where the minus sign in front of ß_+ reverses the slope effect produced by that term.

4.3.6 Selecting the positive feedback components

The choice of the phase compensation components in Fig. 4.17 closely follows that of the negative feedback case of Sec. 4.3.3. As shown in Fig. 4.18, the component selection again redirects the $1/ß$ intercept with A_{OL}. At that intercept, the high-frequency level of $1/ß$, $1 + R_2/R_C$, intercepts A_{OL} at the amplifier's minimum stable gain A_{min} for optimum bandwidth. As described before, A_{min} generally occurs at the second amplifier response pole f_{p2}, and placing the intercept there makes $f_i = f_{p2}$. These conditions duplicate those of Sec. 4.3.3, with the exception that $1/ß_0 = 1$ for the voltage follower considered here. Making this one change adapts the previous results to Fig. 4.17, defining the choice for R_C as

$$R_C = \frac{R_2}{A_{min} - 1} = \frac{R_2}{f_c/f_{p2} - 1}$$

Next the selection of C_C places the pole of the $1/ß$ response a decade below the f_i intercept. This separation assures that the f_p and f_z singularities shown produce canceling phase effects at the intercept. For $f_p = 1/2\pi R_C C_C = 0.1 f_i$, the compensation capacitor becomes $C_C = 5/\pi R_C f_i$. From before, $f_i = f_{p2}$, making $C_C = 5/\pi R_C f_{p2}$. Alternately, $A_{min} = f_c/f_{p2}$, as shown in Sec. 4.3.3, for $f_{p2} = f_c/A_{min}$. Accommodating either case,

$$C_C = 5/\pi R_C f_{p2} = 5A_{min}/\pi R_C f_c$$

These above expressions for R_C and C_C define the relative values of the compensation elements. However, the absolute values depend upon first choosing a value for R_2. Previously, this resistor was part of the negative feedback network and was chosen with other criteria. However, in Fig. 4.17, R_2 only serves as a phase compensation element, and its resistance value remains free to be set. In this choice, the circuit's input impedance and noise performance compete. With $Z_I = A_{OL} Z_C$, the circuit's input impedance increases with higher values of R_C and the accompanying lower values of C_C. This suggests making R_2 large to accommodate a larger resistance for R_C. However, R_2 also generates a voltage noise of $e_{nR2} = \sqrt{4KTR_2}$ right at the amplifier input, making the choice of the R_2 value a compromise.

A logical compromise sets R_2 so that its noise voltage equals one-third that of the inherent amplifier input noise voltage, or $e_{nR2} = e_{ni}/3$. This choice produces the maximum input impedance possible without degrading the noise performance significantly. The one-third factor here turns into one-ninth influence in the rms summation of resistor and amplifier noise voltages. There the rms summation $\sqrt{e_{rR}^2 + e_{ni}^2}$ first raises each noise term to the second power, converting the one-third factor to the one-ninth influence. For this compromise,

$$R_2 = e_{ni}^2/36KT$$

Here, e_{ni} is the amplifier input noise voltage, K is Boltzmann's constant, or 1.38×10^{-23}, and T is temperature in kelvins, or °C + 273.

4.3.7 Compensating the differential input configuration

Differential input connections of op amps impose a special restriction upon external phase compensation. The differential inputs produce high common-mode rejection, but an external phase compensation can degrade this benefit if chosen without regard for the impedance balance requirement of the configuration. Impressing a common-mode voltage across a phase compensation element introduces a signal current that imbalances the circuit's common-mode response. Section 4.2.9 describes the addition of a matching compensation element to restore the balance for one phase compensation method. However, wherever possible, the initial imbalance should simply be avoided by placing the phase compensation elements where they support no common-mode swing. The multipurpose phase compensation methods considered here permit this placement through the choice of the compensation method used.

Figure 4.19 illustrates the problem and its solution for the difference amplifier connection. There the common-mode signal e_{icm} drives the e_1 and e_2 inputs, and under balanced circuit conditions, the difference amplifier rejects the e_{icm} signal, producing no signal at the e_o output terminal. However, a common-mode input signal does produce signals at the op amp inputs, creating a signal difference between these inputs and the op amp output. The voltage divider formed by R_3 and R_4 transmits a portion of the e_2 signal to the op amp noninverting input, and feedback develops a matching signal at the op amp inverting input. This common-mode signal drives any output-to-input connected phase compensation. The dashed connections of the figure illustrate such a case with the positive feedback compensation of the previous section. There the compensation elements draw a common-mode signal current from the junction of R_3 and R_4. The circuit amplifies the resulting voltage to produce an e_o component resulting from the common-mode input signal, and this degrades the circuit's common-mode rejection. Similar degradation results when using the load-decoupling phase compensation described in Sec. 4.1.2.

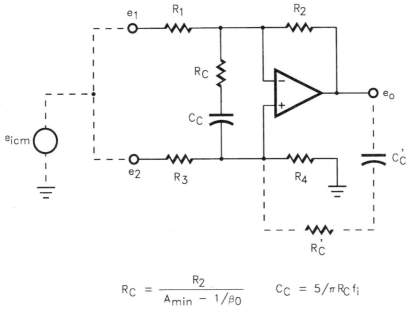

$$R_C = \frac{R_2}{A_{min} - 1/\beta_0} \qquad C_C = 5/\pi R_C f_i$$

Figure 4.19 For differential input connections, Fig. 4.13 type phase compensation isolates compensation elements from common-mode signals

For differential input connections, any external phase compensation should generally be of the configurations shown in Fig. 4.5 or Fig. 4.13. Those configurations avoid common-mode voltage swing across the compensation elements. Only when compensating for input capacitance effects, as in Sec. 4.2.9, should a phase compensation element be connected between an op amp input and output when using a differential input connection. In such a case, the phase compensation also removes a gain-peaking error as described earlier. For more general applications, the pole-zero method of Fig. 4.5 places the compensation elements in series with the amplifier output, avoiding the input-to-output common-mode signal.

Alternately, the method in Fig. 4.13, included in Fig. 4.19 with solid lines, reduces the signal across R_C and C_C to the differential error signal between the op amp inputs. This error signal contains only a fragment of the impressed common-mode signal. The resulting signal current in R_C and C_C still degrades common-mode rejection, but far less than the dashed-line case. Selection of the R_C and C_C follows directly from the discussion of Fig. 4.13. That figure lacks the R_3 and R_4 resistors shown here, but these resistors do not alter the feedback factor. The phase compensation effects remain the same for the two circuits.

References

1. J. Graeme, "Phase Compensation Extends Op Amp Stability and Speed," *EDN*, September 16, 1991.
2. R. Burt and R. Stitt, "Circuit Lowers Photodiode Amplifier Noise," *EDN*, September 1, 1988.
3. J. Dostal, *Operational Amplifiers,* Elsevier, Amsterdam, 1981.
4. J. Graeme, "Phase Compensation Counteracts Op Amp Input Capacitance," *EDN*, January 6, 1994, p. 97.
5. J. Graeme, *Photodiode Amplifiers: Op Amp Solutions,* McGraw-Hill, New York, 1996.
6. J. Graeme, "Capacitor Permits Higher Slew Rates," *Electronic Design,* March 28, 1991.

Chapter 5

Reducing Radiated Interference

Diminishing returns eventually limit the noise reduction achieved through measures focused upon the visible, hard-wired portion of a circuit. As a companion part of every circuit, an invisible but very real portion couples radiated noise into the signal path. There electrostatic and magnetic coupling impose a background noise floor that requires attention to the amplifier's environment instead of the intended circuit. Physical separation and shielding remain the first defense against this coupling[1,2] but op amp circuits present additional noise reduction opportunities through differential inputs, impedance balancing, and loop minimization. An evaluation of the coupling mechanisms guides the application of these techniques.

Limiting the effects of these external noise sources requires attention to the circuit's location, shielding, the circuit's structure, and the physical arrangement of the circuit components. Electrostatic and magnetic noise signals enter the circuit through parasitic mutual capacitances and mutual inductances, respectively. For both effects, maximizing the physical separation of the amplifier from the source first minimizes the coupling parasitics. Next, shielding absorbs both electrostatic and magnetic fields, but the two field types often impose different shield material requirements. In addition, converting an op amp circuit to a differential input configuration adds common-mode rejection to the noise reduction. Simple impedance balancing extends this common-mode-rejection benefit to more general op amp configurations. To reduce magnetic coupling still further, the identification of an op amp circuit's potential receptor loops and the evaluation of the corresponding noise gains define the priorities for loop area minimization. Loop identification also defines those areas in the circuit that potentially produce rather than receive magnetic noise signals.

5.1 Reducing Electrostatic Coupling

Electrostatic or electric field coupling, such as from the power line, supplies noise signals through the mutual capacitances that exist between any two objects. The objects serve as the capacitor's plates, and the intervening air or other medium acts as the capacitor's dielectric layer. Ac voltage differences between the two objects drive these mutual capacitances, coupling noise currents between the two. Ideally, electrostatic shielding intercepts these currents and shunts them to ground. Alternately, differential-input or impedance-balanced configurations make the coupled signal a common-mode effect for removal by the op amp's common-mode rejection.

5.1.1 Electrostatic shielding

Shield material and grounding determine an electrostatic shield's effectiveness. Using a shield material with high electrical conductivity ensures that the coupled currents produce little voltage drop across the shield. Then little of the original field continues within the shield boundaries. To be effective, the shield's ground must be earth ground because only this remains a common reference for the separate objects involved in the coupling.

Further, the shield should be connected to the system ground to minimize the effects of the parasitic capacitances introduced. All shield enclosures form parasitic capacitances with each component shielded. When left ungrounded, the shield develops a signal voltage largely determined by mutual capacitance coupling from the circuit's larger signals to the shield. Any resulting shield voltage couples back into the circuit through other mutual capacitances. Returning the shield connection to the system common removes the shield signal, relative to the circuit, and that removes the associated self-coupling. For example, shield-conducted capacitive currents from the output of an amplifier then flow to ground rather than creating a shield voltage signal that could couple into sensitive circuit areas. This grounding also avoids a bandwidth restriction potentially imposed by a shield-induced parasitic capacitance that would bypass the amplifier's feedback.

5.1.2 Common-mode rejection of electrostatic coupling

Op amp common-mode rejection also reduces the effects of electrostatic coupling, as an alternative to shielding or to remove the residual effects resulting from shield imperfections. Typically, electrostatic noise interference couples to all points of a circuit nearly equally, making this interference a natural candidate for removal by an op amp's common-mode rejection. However, most op amp circuit configurations fail to utilize the amplifier's common-mode-rejection capability. In those cases, an impedance imbalance at the op amp's two inputs commonly disables this feature. Switching to a differential input connection or balancing these impedances activates common-mode rejection for reduced coupling sensitivity and reduced dc error as well. However, op amp common-mode rejection does not generally replace shielding because it declines with increasing frequency and the electrostatic coupling to the two amplifier inputs will not be exactly the same. A combination of shielding and common-mode rejection produces the best overall result.

Figure 5.1 models the basic electrostatic coupling effect with an inverting op amp configuration, an electrostatic noise source e_e, and a mutual capacitance C_M. Source e_e represents any ac voltage that creates an electric field in the vicinity of the amplifier. This voltage couples a noise current i_{ne}, through C_M and into the amplifier's feedback network. None of that current flows into the feedback resistor R_1 because that would change the voltage at the op amp's inverting input. The circuit's feedback action inherently forces this voltage to equal that at the op amp's noninverting input, zero in this case. To do so, feedback develops a voltage on R_2 that accommodates the flow of i_{ne} by producing an output noise signal equal to $-i_{ne}R_2$. In practice, other mutual capacitances couple noise currents to other points in the circuit as well, but the low impedances there usually minimize the effects of those currents.

Switching to a differential input configuration produces balanced conditions that reject the $-i_{ne}R_2$ noise signal through the op amp's

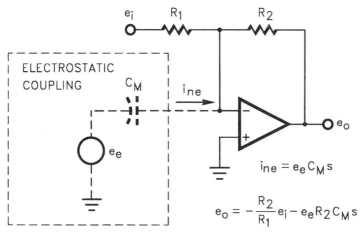

Figure 5.1 Electric fields couple noise current through parasitic mutual capacitance into op amp feedback networks.

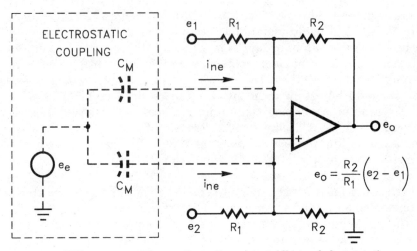

Figure 5.2 Difference amplifier configurations reject effects of electrostatic coupling by developing a canceling signal and through op amp common-mode rejection.

common-mode rejection. Figure 5.2 models this case with the difference amplifier configuration and two mutual capacitances coupling to the op amp's two inputs. These capacitances match as long as the two inputs reside at essentially equal distances from the noise source. In addition, the voltages on the capacitances match since the amplifier's feedback forces the two inputs of the op amp to the same voltage. Thus the two capacitors couple equal i_{ne} noise currents to the circuit's two resistor networks, and these currents produce canceling effects. The current coupled to the op amp's noninverting input produces a noise voltage of $i_{ne}(R_1 || R_2)$. Then the circuit amplifies this voltage with a gain of $1 + R_2/R_1$ to produce an output component of noise equal to $i_{ne}R_2$. As in the previous inverting case, the i_{ne} current coupled to the op amp's inverting input flows through the R_2 feedback resistor, developing an output component of noise equal to $-i_{ne}R_2$. Thus the two i_{ne} noise currents here produce equal and opposite output noise components, so their effects cancel. Only the $i_{ne}(R_1 || R_2)$ signal effect remains at the two op amp inputs, as a common-mode signal, and the amplifier's common-mode rejection isolates the output from this effect.

5.1.3 Common-mode rejection with nondifferential circuits

The circuit operation described for the difference amplifier suggests a simple noise-coupling remedy for more general op amp configurations. In the preceding analysis, the $i_{ne}(R_1 || R_2)$ signal at the op amp's

Reducing Radiated Interference 159

inputs remains the same when an appropriate balancing resistor R_B replaces the lower R_1 and R_2 resistors. Then making $R_B = R_1 || R_2$ retains the noise cancellation described and permits simple modifications to other op amp configurations for similar electrostatic noise reductions. Figure 5.3 shows this solution for the typical inverting and noninverting configurations. There simply adding the R_B balancing resistances in series with the op amps' noninverting inputs capitalizes upon the noise currents coupled there to produce the noise cancellation.

Further examination reveals the impedance balance produced by this solution and a side benefit to offset reduction. In the figure, mak-

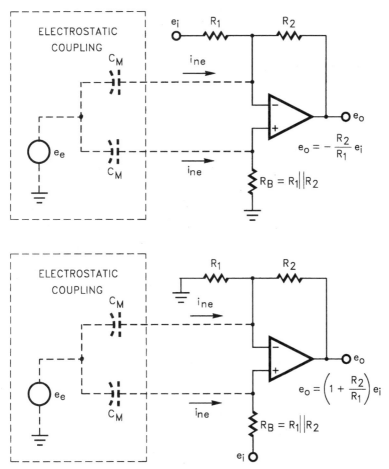

Figure 5.3 Adding impedance balancing resistors in series with the op amp's noninverting input extends noise rejection of difference amplifiers to more general op amp configurations.

ing $R_B = R_1||R_2$ as prescribed defines the resistance level at the op amp's noninverting input. From the figure, the impedance seen from the op amp's inverting input also equals $R_1||R_2$. There R_1 returns to the low impedance of a source and R_2 returns to the low impedance of the amplifier output. Thus for impedance analysis purposes, the latter two resistors effectively return to ground and appear in parallel. Then the circuit's resistors present equal impedances at the op amp's two inputs and common-mode rejection will reject electrostatic coupling.

For both configurations illustrated here, the R_B resistors resemble those often added to reduce the dc offset produced by an op amp's input bias currents. For that purpose, adding a resistance equal to $R_B = R_1||R_2$ in series with the op amp's noninverting input commonly produces an associated offset cancellation. When applied in those cases, a capacitive bypass of the R_B resistance avoids two side effects of this offset solution. This bypass limits the added resistance noise effect and avoids the low-pass filter formed by the resistance with the op amp's input capacitance. In this case, the R_B resistors also produce this offset cancellation, but the desired noise coupling cancellation precludes the bypass benefits. Such bypass would again imbalance the impedances seen from the op amp's two inputs, preventing common-mode rejection reductions of coupled noise at higher frequencies. As a result, the noise of the R_B resistance adds to the circuit's noise response for both cases illustrated. Also for the noninverting configuration, elimination of the bypass allows resistor R_B to form a low-pass filter with the op amp's common-mode input capacitance, potentially restricting the circuit's bandwidth.

The effectiveness of this noise cancellation depends upon precise resistance matching and upon high op amp common-mode rejection. At lower frequencies, the resistance error of the required $R_B = R_1||R_2$ condition typically dominates and limits the cancellation accuracy to about the same error level. At higher frequencies, the response roll off of the op amp's common-mode rejection increases this error in proportion to 1/CMRR.

5.2 Reducing Magnetic and RFI Coupling

Magnetic noise coupling and radio-frequency interference (RFI) introduce circuit noise through a common coupling mechanism, mutual inductance. There the interference source acts like the primary of a transformer and circuit loops act like secondary windings. While often considered separately, RFI simply represents the higher-frequency form of parasitic magnetic coupling. However, in shielding, this frequency distinction suggests the separation because the frequency of the magnetic source greatly influences the effectiveness of magnetic shielding materials. At lower frequencies, only the magnetic property of ferrous metals permits a practical shield thickness. However, at higher frequencies, decreases in both the magnetic response and the shield thickness requirement make copper a good alternative material. There even the copper layer of a ground plane becomes effective. In addition to shielding, carefully choosing the amplifier's physical and electrical configurations reduces magnetic coupling effects. Minimizing the physical areas of circuit loops reduces the coupling efficiency, and using a differential input structure makes common-mode rejection reduce the residual effect.

5.2.1 Magnetic shielding

A comparison of magnetic and electrostatic coupling and a transformer analogy illustrate the added shielding requirement of the magnetic case. The electrostatic shielding described before simply requires a shield material of high electrical conductivity. There the high conductivity short-circuits the currents transmitted through mutual capacitances to ground. However, magnetic fields couple through mutual inductances, rather than capacitances, and typically produce voltage rather than current signals in op amp circuits. The coupled signals develop in all circuit loops within the field and produce relative coupling magnitudes determined by the loop areas. For the magnetic field, a high-conductivity electrostatic shield only forces an equipotential condition at the shield boundaries. Grounding this shield does not terminate the magnetic field and only establishes a zero voltage reference. By analogy, grounding the center tap of a transformer's secondary winding establishes a reference voltage, but does not terminate the transformer's coupling. Thus an electrostatic shield distorts the magnetic field but does not necessarily remove the field energy, and some portion of the field's energy continues the field within the shield boundaries.

Fortunately for noise reduction, some of the magnetic field energy dissipates in eddy currents and ohmic absorption within the shield. Shield material selection optimizes this dissipation with different results for low and high frequencies. To reduce magnetic coupling effectively, the shield must absorb the field energy as the field travels through the shield's walls, and a combination of high conductivity and high magnetic permeability produces the greatest field absorption. High electrical conductivity ensures that the induced shield currents produce little voltage drop across the shield, preventing continuation of the field through electrostatic coupling within. High magnetic permeability ensures efficient absorption of the magnetic field moving through the shield. At the power-line frequency, common to many magnetic coupling effects, the magnetic permeability of ferrous materials reduces magnetic coupling an order of magnitude better than other shield materials.[4] For such materials, the realignment of magnetic dipoles in the ferrous material consumes magnetic field energy in the form of eddy currents. This property greatly improves a ferrous magnetic shield's effectiveness in spite of the fact that the electrical conductivity of steel remains about an order of magnitude lower than that of copper. However, at power-line frequencies, the shield thickness required for a copper shield becomes prohibitive due the great skin depth of this material.

At radio frequencies, copper becomes a more reasonable, but never superior, alternative shield material. There both the magnetic permeability of ferrous metals and the skin depths of all metals drop dramatically. Above 10 kHz, the magnetic permeability of ferrous metals drops due to the finite time required to realign the material's magnetic dipoles. The shorter periods of high-frequency signals preclude the realignment and prevent the resulting conversion of field energy to eddy currents. Then a shield must remove magnetic field energy through ohmic absorption, as reflected by skin depth. Skin depth indicates the thickness of a given shield material required to attenuate a magnetic field by a factor of $e = 2.73$. Fortunately, skin depth also decreases at higher frequencies, due to the shorter wavelength of the signal. Thus at high frequencies, the decreased permeability of ferrous materials reduces their relative advantage over copper, and the reduced skin depths associated with the signal frequency make the required copper thickness more practical. However, for a given shield thickness, steel still retains about a factor of 3 absorption advantage.

5.2.2 Minimizing loop areas

As described, physically separating and shielding an amplifier from a magnetic noise source offers the best protection against magnetic noise coupling. However, the amplifier's physical and electrical configurations also affect this coupling. Configuring the amplifier's layout for minimum physical loop areas and designing the circuit for common-mode rejection both reduce the resulting noise. Minimizing the areas of the circuit's loops minimizes the mutual inductances that couple the magnetic noise signal. Careful component layout generally achieves this by simply placing the circuit's components close to the op amp and then minimizing the lengths of the component interconnections. Leadless chip components and smaller op amp packages aid in this area reduction. In addition, an op amp's common-mode rejection reduces magnetic coupling effects when using a differential input

Reducing Radiated Interference   163

configuration. For that configuration, matching loop areas and distances from a noise source makes the amplifier reject the effects of some of the coupled noise signals.

For op amp circuits, the most confusing task in magnetic coupling reduction can be the identification of the receptor loops. The physical arrangement of the circuit's components forms these loops in several ways, and Fig. 5.4 shows the loops produced by a noninverting op amp circuit. There the op amp forms three loops through its connections with the source, the feedback, and the load. The op amp would seem to break these loops but, as will be seen, the amplifier's feedback action continues them. Differing amplifier noise responses and grounding connections differentiate the three loops shown. First, the ground return of the R_1 resistor forms loop L_1 with coupled noise source e_{m1}, the signal source, and the op amp input circuit. It might seem that the amplifier's very high input impedance would break the loop by interrupting an otherwise continuous conductive path.

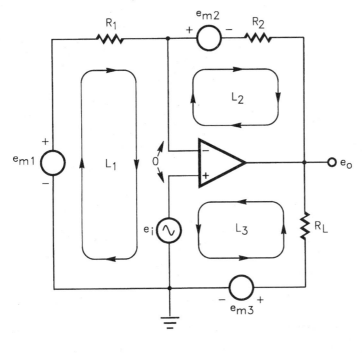

$$e_o = \left(1 + \frac{R_2}{R_1}\right)e_i - \frac{R_2}{R_1}e_{m1} - e_{m2}$$

Figure 5.4 Op amps form magnetic receptor loops with external components and couple magnetic field noise to the circuit output.

However, feedback forces the voltage between the amplifier inputs to zero, just as if a short circuit connected them, and this completes the L_1 loop. It might also seem that the ground connection in this loop would produce an interruption. However, it does not, just as grounding a center tap on a transformer secondary does not interrupt the transformer's coupling action. This loop's noise signal e_{m1} drives an inverting amplifier configuration and receives a gain of $-R_2/R_1$. This gain potentially makes the L_1 loop a serious noise source and a priority choice for area minimization.

Next consider the L_3 loop and its coupled signal e_{m3}. Less obvious, this loop results from connecting e_i and R_L together at the circuit's ground return and from a signal path through the amplifier. Feedback action again continues the loop through the amplifier. Signal e_i drives the input of a noninverting amplifier, and feedback makes the amplifier output respond to e_i, maintaining loop continuity between the top sides of e_i and R_L. The resulting e_{m3} noise signal transfers to the load with unity gain.

The final circuit loop, L_2, depends upon the loop continuity conditions described for L_1 and L_3. As described, feedback action maintains continuity from the amplifier's noninverting input to both the inverting input and the output. Visual examination of Fig. 5.4 shows that making these two effective connections completes the L_2 loop as well. The resulting e_{m2} noise signal also transfers to the circuit output with unity gain.

5.2.3 Common-mode rejection of magnetic coupling

Minimizing the preceding loop areas reduces but does not eliminate the noise signal coupled from magnetic sources. Some finite loop areas always remain to receive this coupling. However, the difference amplifier offers further noise reduction through an op amp's common-mode rejection. Unlike the electrostatic case, this circuit configuration fails to reject the coupled noise of the primary input loop but can produce canceling effects in the secondary loops of the circuit's feedback.

For the difference amplifier, Fig. 5.5 shows the relevant loops and their coupled noise signals. There the ground of the differential input acts as a center tap for the L_1 receptor loop, splitting the e_{m1} signal into two equal parts. Unfortunately, those two parts present opposite-polarity signals to the R_1 resistors of the differential input circuit. As a result, the net e_{m1} presents a differential rather than a common-mode input signal, and the op amp's common-mode rejection does not reject this signal.

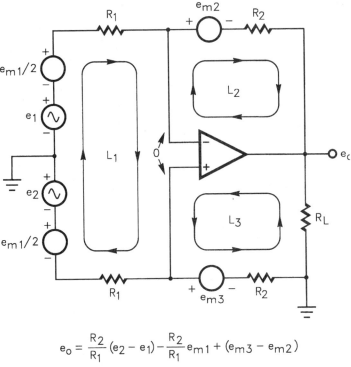

$$e_o = \frac{R_2}{R_1}(e_2 - e_1) - \frac{R_2}{R_1}e_{m1} + (e_{m3} - e_{m2})$$

Figure 5.5 Difference amplifiers fail to reject the e_{m1} noise signal of the L_1 loop but, with loop matching, make e_{m3} and e_{m2} effects cancel.

However, the balanced structure of the difference amplifier still permits coupled noise reduction through matching of the L_2 and L_3 loops. These loops produce the e_{m2} and e_{m3} signals that tend to produce canceling effects at the circuit's output. First, signal e_{m3} develops a signal equal to $e_{m3}R_1/(R_1 + R_2)$ at the op amp's noninverting input. This represents both a common-mode signal to the op amp and a noninverting input signal to the circuit. Feedback replicates this noninverting input signal at the op amp's inverting input to make this a common-mode signal, and the op amp's inherent common-mode rejection isolates the circuit's output from this effect. This same signal also excites the circuit as a noninverting amplifier but, with loop matching, the impedance balance that activates common-mode rejection for the difference amplifier configuration also rejects this effect. First, the noninverting input signal $e_{m3}R_1/(R_1 + R_2)$ noise receives a voltage gain defined by the upper resistor network of the figure. The impedance balance of the difference amplifier makes this gain a fortuitous $(R_1 + R_2)/R_1$ to reduce the resulting output signal to simply e_{m3}.

Counteracting this noise effect, signal e_{m2} produces an output noise component of opposite polarity from the same noise source. Signal e_{m2} cannot alter the voltage at the op amp's inverting input because feedback forces this input voltage to equal that at this amplifier's noninverting input. Thus e_{m2} produces no signal voltage across the upper R_1 input resistor and no resulting feedback current. Without such current, the upper R_2 resistor likewise produces no associated output signal. Then e_{m2} transfers directly the op amp's output with unity gain, and observing polarity, this effect produces an output signal component equal to $-e_{m2}$. Together the e_{m2} and e_{m3} noise components produce an output noise signal equal to $e_{m2}-e_{m3}$, and making $e_{m2} = e_{m3}$ produces noise cancellation of these two effects. While not immediately apparent from this lengthy discussion, a reexamination of the analysis steps described shows that this cancellation results from the same impedance balance condition that produces the fundamental common-mode rejection of the difference amplifier configuration.

This cancellation requires matching the physical areas of the L_2 and L_3 loops, their distances from any interfering magnetic source, and their orientations relative to that source. Matching these three features equalizes the magnitudes and phases of e_{m2} and e_{m3} for accurate cancellation. Matching loop areas and distances equalizes the magnitudes of e_{m2} and e_{m3}, and the matching distances also produce a first-order phase equalization. Accurate phase matching, as required for higher-level common-mode cancellation, also requires matched loop orientations relative to the magnetic source. Most often, this noise reduction opportunity aids in the rejection of low-frequency, local noise sources such as power transformer interference. There the long wavelengths of the low frequencies involved reduce the precision required in orientation matching.

For higher-frequency interference, RFI filtering somewhat replaces common-mode rejection to counteract the accompanying increase in magnetic coupling efficiency. This increase potentially dominates noise performance, often due to the RFI of coresident digital circuitry. However, the bandwidth limitation of the amplifier permits a filtering solution. The limited bandwidth of the op amp frequently restricts the circuit's response to frequencies well below the radio-frequency range. This permits filter removal of RFI signals following the amplifier without restricting the circuit's useful bandwidth. However, output filtering does not remove a potentially debilitating side effect of higher-level RFI. In that case, the input circuit of an op amp acts like a radio-frequency detector, separating a lower-frequency envelope from a carrier.[3] There larger-level radio-frequency signals drive the emitter-base junctions of bipolar input transistors to produce a rectifying action, and the transistor's junction capacitances store the envelope level of the radio-frequency signal at the amplifier input. If not totally disabled by this interference, the amplifier at least transmits an amplified replica of the envelope to the circuit output. Field-effect-transistor (FET) input op amps significantly reduce the likelihood of this effect due to the larger voltage swing required to produce a rectifying action.

5.3 Reducing Multiple Coupled-Noise Effects

Coupled noise seldom provides clear clues as to its origin, and successful reduction of this noise requires a modified troubleshooting approach. In a given application, any one of the preceding noise reduction techniques may offer the best solution, or the application may require a combination of two or more. Without definitive clues, a trial-and-error application of the noise reduction practices prevails. However, the combinational possibility requires a modification to this approach. Traditionally, when a given trial implementation fails to solve the problem, you remove this implementation and proceed to the next. However, with the potential for combinational effects, effective noise coupling reduction requires a delay in this removal. When the trial of a first noise reduction technique fails to solve the problem, do not remove it before applying other techniques. The effect of the first technique may be masked by the requirement for others, and removing the first at this point may suboptimize the final solution. Instead, proceed through the series of noise reduction practices described, leaving all implementations in place until the end. Then with the coupling reduction optimized, begin to remove those implementations which originally appeared to have no effect. In the reduced-noise background of the optimized condition, the actual noise effects of the individual implementations become apparent. Remove the ineffective ones at this point but not before.

5.4 Minimizing Magnetic Field Generation

The preceding discussions focus upon a circuit's sensitivity to coupled noise, overlooking the circuit's potential for producing such noise. Simple current flow in a conductor produces a magnetic field capable of coupling noise back into the amplifier or to other circuitry. The current flows in op amp circuits typically lack the magnitudes needed to produce a significant field, except at the amplifier output. There the load current supplied by the amplifier produces a potentially significant field. However, layout attention to coaxial returns minimizes that significance.

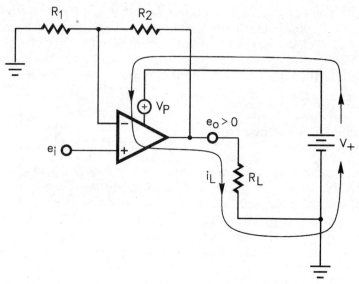

Figure 5.6 Op amp load current flows in a loop capable of producing magnetic interference.

As illustrated in Fig. 5.6, the amplifier supplies the load current i_L to load resistance R_L, often producing a significant current flow in connecting conductors. A closer examination of this current's path reveals current flow in a field-generating loop. For the example shown, the amplifier produces a positive output voltage e_o by drawing the load current i_L from the V_+ positive supply and delivering it to R_L. Then R_L's ground connection returns the current to V_+, completing the loop. A similar current conduction loop results for negative e_o values through the amplifier's negative power supply. In each case, the magnitude and frequency content of i_L and the physical area of the conduction loop determine the strength of the magnetic field generated. Application requirements dictate the characteristics of i_L, but physical layout options permit minimizing the corresponding loop area to greatly reduce the resulting field strength.

Coaxial conduction of the supply and return of the i_L current minimize this loop area. For the simplest illustration of the coaxial principle, Fig. 5.7 shows this current conduction with a coaxial cable that conducts the supply and return of i_L. In practical cases, simpler options suffice for this current conduction, but this example demonstrates the coaxial return principle. A later discussion presents these options using the principle illustrated here. In the figure, the cable's center conductor supplies the i_L current to the amplifier and the cable shield returns it to the power-supply common. Just the appearance of

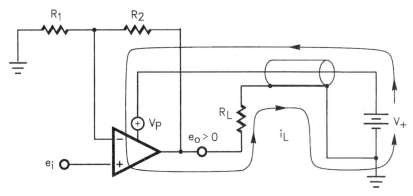

Figure 5.7 Coaxial return of op amp load current minimizes loop area and associated magnetic radiation.

the resulting figure communicates the reduced area of the conduction loop, supplying an intuitive sign of magnetic field reduction. In practice, the relative area reduction far exceeds that presented by the short cable length of the illustration.

The significance of this area reduction becomes apparent when considering the magnetic fields produced by the supply and return conductors. To minimize the loop area, these conductors must follow essentially the same physical path. Then the two conductors carry the i_L current in opposite directions along this common path, producing equal magnetic fields of opposite orientation. Such fields tend to cancel each other, neutralizing the net magnetic field. The coaxial cable optimizes this canceling effect by producing a common centroid or axis for the two current flows. While difficult to illustrate, the return current conducted in the cable shield of Fig. 5.7 distributes throughout the shield, enveloping the center conductor. Averaged around the circular shield, the net magnetic field produced has a centroid coincident with the center conductor. Thus the supply and return conductors of the coaxial cable permit precise magnetic field neutralization. However, coaxial cable connections represent an impractical solution to the multitude of current conduction loops found on a typical circuit board.

Fortunately, adjacent board traces or a ground plane provide nearly coaxial performance for the practical case. Adjacent circuit-board traces illustrate this solution but the ground plane supplies it best. Conceptually adjacent traces, supplying and returning a given current, approximate the coaxial condition. This configuration again minimizes the loop area and only slightly compromises the ideal coaxial conduction of the cable. With adjacent traces, the space separating the two differentiates the central axes of the two magnetic fields produced, slightly reducing the magnetic field cancellation. Still, the adjacent trace solution also remains impractical for all but the simplest of circuit boards. Layout implementation of a return trace adjacent to each signal trace very soon encounters conflicting layout demands.

In practice, a ground plane avoids the adjacent trace complication and provides similar coaxial performance. There serendipity assures the coaxial return at the higher frequencies where magnetic coupling becomes increasingly significant.[4] A return current delivered to the ground plane follows the path of least impedance back to the power-supply common. At lower frequencies, the ground plane resistance controls this impedance and the return current typically follows the shortest physical path to the supply common. This does not establish a coaxial return, but the signal's lower frequency prevents the production of a noise-significant magnetic field. At higher frequencies, the ground plane inductance controls this return impedance and automatically assures a coaxial return. There the loop area of the conduction path corresponds to inductance, and the loop's magnetic field controls the path of the return current flow. Seeking the path of least impedance, the return current follows a path in the ground plane paralleling that of the corresponding supply current in the circuit plane. This current flow combination minimizes the return path inductance and the associated magnetic field. The physical separation between the circuit plane and the ground plane still differentiates the axes of the two conduction paths, but only by the thickness of the circuit board.

References

1. R. Morrison, *Grounding and Shielding Techniques in Instrumentation*, 2d ed., Wiley, New York, 1977.
2. H. Ott, *Noise Reduction Techniques in Electronic Systems,* Wiley, New York, 1976.
3. Y. Sutu and J. Whalen, "Statistics for Demodulation RFI in Operational Amplifiers," presented at the IEEE International Symposium on Electromagnetic Compatibility, August 23, 1983.
4. P. Browkaw and J. Barrow, "Grounding for Low- and High-Frequency Circuits," *Analog Dialog,* vol. 23, no. 3, 1989.

Chapter

6

Distortion and Its Measurement

Chapter 1 develops fundamental op amp performance analysis and encompasses most op amp error sources, but it does not specifically address distortion. In practice, the same analysis techniques apply to the distortion error once the nature of op amp distortion is defined. This chapter examines this distortion, models the effect, and describes measurement techniques that exploit the unique nature of the op amp distortion signal.

Many characteristics of an op amp circuit introduce distortion, but feedback consolidates their effects in the amplifier's input error signal. This signal contains all of the amplifier's distortion products in the background of only a small fraction of a test signal. The consolidation makes op amp distortion an input-referred characteristic just like offset voltage, and two conveniences result from this input-referred nature. First, the input-referred characteristic measured in a given application configuration permits prediction of the output-referred distortion for any other configuration. This convenience parallels that of an op amp's input offset voltage, where the measurement of the input-referred offset permits prediction of the output-referred result for any circuit configuration. In the second convenience, feedback separates the op amp distortion component from the test signal by producing an input error signal that represents the circuit's feedback error. This separation increases measurement resolution dramatically by avoiding two fundamental test equipment limitations. First, the test signal removal also removes the distortion products of that signal, avoiding one measurement resolution limit. Second, this removal reduces the background signal of the measurement, avoiding the dynamic range limit of the signal analyzer.

Three fundamental measurement techniques exploit the distortion signal separation of an op amp. Direct measurement, selective amplification, and bootstrap isolation all focus upon the separated input error signal to improve measurement resolution. In the simplest case, direct measurement of the feedback error signal yields the amplifier's input-referred distortion. However, this approach often requires the addition of a measurement amplifier, and this amplifier's distortion sets a new resolution limit. To avoid that limit, the other two measurement techniques avoid the added amplifier. With the next technique, the amplifier under test selectively amplifies the input error signal, replacing the measurement amplifier without altering the measurement result. This technique magnifies the relative significance of the distortion signal, but it also reintroduces the test signal in the measurement. With the final measurement technique, bootstrap isolation of the feedback error signal removes both the added amplifier and the test signal. Each of the three measurement techniques offers a different set of advantages for different op amp circuit configurations. In some cases, a combination of two techniques offers the greatest measurement resolution.

6.1 The Nature of Op Amp Distortion

Numerous practical limitations introduce distortion in the transmission of a signal through an op amp. The nonlinear transfer characteristics of transistors, the voltage coefficients of junction capacitances, the hyperbolic responses of differential stages, thermal feedback, and other effects produce distortion. In spite of the numerous distortion sources, an op amp's feedback consolidates all of their effects, easing the associated analysis and measurement. Feedback isolates the op amp's distortion products in the amplifier's input error signal, where the distortion signal remains relatively independent of the circuit's feedback connection. This independence makes op amp distortion an input-referred characteristic, and knowing this input-referred error for a given op amp permits the prediction of output distortion for any feedback connection of that amplifier.

In the discussion that follows, the examination of three basic feedback connections first demonstrates the input-referred nature of the distortion signal. Then a closer examination of the distortion signal isolates its sources to the nonlinearities of open-loop gain and common-mode rejection. These amplifier characteristics define input error signals that vary with signal levels but remain independent of the application circuit's feedback connection. The latter independence establishes the input-referred nature of op amp distortion, reducing characterization and analysis tasks.

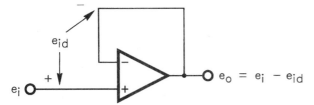

Figure 6.1 For voltage followers, feedback consolidates all differences between input and output signals in error signal e_{id}.

6.1.1 The Op amp distortion signal

The voltage follower of Fig. 6.1 first demonstrates the input location of the distortion signal. There the applied input signal e_i drives the op amp input, producing an approximately equal output signal e_o. Only the op amp's differential input error signal e_{id} differentiates the input and output signals. For this circuit, a simple loop equation locates any distortion introduced by the op amp through

$$e_o = e_i - e_{id}$$

As trivial as this equation seems, it displays the fundamental location of op amp distortion signals. Distortion introduced by the op amp produces a difference between the follower's input and output signals. This equation states that the output signal e_o remains a replica of the input signal e_i except for the input error signal e_{id}. Thus e_{id} must include any distortion products introduced by the amplifier.[1] Recognition of this fact simplifies distortion understanding and measurement for op amp circuits in general.

Only the voltage follower presents this simple equation to isolate the distortion signal. However, the underlying principle extends to all op amp configurations. Other configurations add feedback networks that alter the input-output relationship from the simple, unity gain of the follower, producing higher gains or inverted gain polarities. However, this linear feedback does not alter the e_{id} representation of op amp distortion products.

The feedback network of the noninverting configuration produces higher gain in Fig. 6.2. This network also interrupts the follower's simple relationship between the e_i, e_o, and e_{id} signals. However, the amplified output signal still compares the same three signals. The gain of the noninverting amplifier simply amplifies the effect of e_{id}, increasing the distortion component in e_o. Here e_{id} subtracts from the signal developed by e_i on R_1, and the resulting R_1 current flows into R_2 to develop an amplified output voltage. The voltage developed on

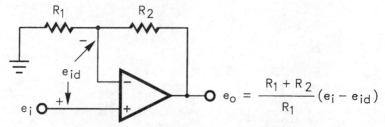

Figure 6.2 Noninverting configurations add gain to the follower case, but still confine all distortion to e_{id}.

R_2 also reflects the effect of e_{id}, amplifying the output distortion result. This amplified distortion remains based upon e_{id}, as seen from the circuit's transfer response equation

$$e_o = \left(1 + \frac{R_2}{R_1}\right)(e_i - e_{id})$$

This equation states that e_o remains an amplified replica of e_i except for a similarly amplified e_{id}. Any distortion introduced by the amplifier still originates in e_{id}, independent of the feedback. Later Sec. 6.1.2 shows that the signal e_{id} itself also remains independent of the feedback connection.

Examination of the inverting configuration produces the same result, as demonstrated with superposition analysis using Fig. 6.3. There a feedback network again alters the simple relationship between the e_i, e_o, and e_{id} signals. Setting e_{id} to zero, superposition analysis defines the relationship between e_o and e_i as the circuit's ideal inverting gain $-R_2/R_1$. Next, setting e_i to zero effectively grounds the input side of R_1. Then the circuit appears as a noninverting amplifier to the signal $-e_{id}$. This signal receives the same 1 +

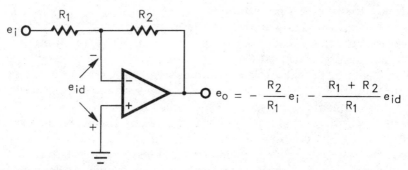

Figure 6.3 Inverting configurations produce different gains for e_i and e_{id}, but again confine all distortion to e_{id}.

R_2/R_1 gain as in the previous noninverting case, and combining the two superposition results produces

$$e_o = -\frac{R_2}{R_1} e_i - \left(1 + \frac{R_2}{R_1}\right) e_{id}$$

Once again, the response equation states that e_o is an amplified replica of e_i except for an amplified e_{id}. Except for the effects of linear amplification, all differences between the input and output signals originate in e_{id}. Thus e_{id} again contains all of the amplifier's distortion products, and this condition repeats for all op amp configurations. For any configuration, the derivation of the circuit's transfer response locates the op amp distortion source in the input error signal e_{id}.

6.1.2 Input-referred op amp distortion

Further analysis confirms op amp distortion as an input-referred characteristic, independent of the feedback network. This greatly eases amplifier characterization and subsequent circuit analysis. The characterization of distortion performance under only one feedback condition defines the input-referred characteristic. Then simple analysis extends the input-referred result to any feedback connection having the same signal and load conditions. The distortion result does vary with those conditions, but later measurement techniques adapt for these variables.

A simplification of the error analysis in Sec. 1.1 isolates the op amp's distortion to feedback-independent amplifier characteristics. That section shows the components of e_{id} to be composed of errors due to input offset voltage, input noise voltage, open-loop gain, common-mode rejection, and power-supply rejection. Knowing that e_{id} contains the distortion error, examination of each e_{id} component identifies those responsible for distortion. To introduce distortion, an amplifier characteristic must first produce an output signal in response to an input signal. This rules out the first two components of e_{id}, the input offset and noise voltages, because these remain fixed errors unaffected by the input signal. Power-supply rejection introduces a signal-dependent error, but in response to power-supply, rather than input, signals. This leaves open-loop gain and common-mode rejection as the only e_{id} sources capable of introducing distortion.

Logical reasoning supports this conclusion. Only input-output transfer responses can introduce distortion in the transmission of a signal. To introduce distortion, the input signal must exercise a source of nonlinearity that, in turn, produces an output response. For most amplifiers, the internal amplifier gain represents the only input-output transfer characteristic. Op amps introduce a second such characteristic in common-mode rejection. Common-mode signals at an op amp input produce an output error signal, reflecting a second transfer response. Nonlinearities in these two transfer characteristics produce the op amp's distortion signal.

Both A_{OL} and CMRR introduce input-referred error signals, as described and analyzed in Sec. 1.1. For distortion analysis, the model used there simplifies to Fig. 6.4. In this figure, a first component of e_{id} results from the finite open-loop gain A_{OL} of the op amp. To support an output voltage of e_o, feedback develops the error signal e_o/A_{OL} between the amplifier inputs. This signal includes a distortion component due to the amplifier's gain variation with signal amplitude, and this variation reflects the associated effects of all amplifier stages. Similarly, the op amp requires a differential input signal to support the output signal resulting from finite CMRR. Analysis quantifies this input signal in terms of the op amp's common-mode input signal e_{icm} and the amplifier's common-mode rejection ratio CMRR. The common-mode rejection ratio is defined as CMRR = A_D/A_{CM}, where A_D and A_{CM} are the differential and common-mode gains of the amplifier. For an op amp, $A_D = A_{OL}$, and solving for the common-mode gain yields $A_{CM} = A_{OL}/\text{CMRR}$. However, by definition, $A_{CM} = e_{ocm}/e_{icm}$, where e_{ocm} is the output signal resulting from an input common-mode signal e_{icm}. Equating the two A_{CM} expressions and solving for e_{ocm} yields $e_{ocm} = A_{OL}e_{icm}/\text{CMRR}$. To support e_{ocm}, feed-

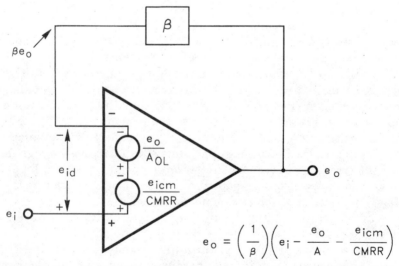

Figure 6.4 Examination of e_{id} reveals the sources of op amp distortion in A_{OL} and CMRR errors.

back develops the gain error signal e_{ocm}/A_{OL} between the op amp's inputs. Substituting the previous e_{ocm} expression defines this error signal as $e_{icm}/CMRR$, so the op amp simply attenuates the input common-mode signal by a factor equal to the amplifier's common-mode rejection. The resulting $e_{icm}/CMRR$ signal includes distortion components produced by signal-dependent variations of both A_{OL} and A_{CM}. These variations reflect to the $e_{icm}/CMRR$ error signal shown through $CMRR = A_{OL}/A_{CM}$.

Analysis of the model in Fig. 6.4 shows op amp distortion to be an input-referred characteristic through the result

$$e_o = (1/\beta)(e_i - e_{id})$$

Thus the model supplies a gain of $1/\beta$ to e_{id} and the included amplifier distortion products. Output distortion varies in accordance with the closed-loop feedback applied by ß. Different op amp applications establish different ß feedback factors, producing different output distortion results. However, at the model's input, e_{id} and the included amplifier distortion remain unchanged by ß. Thus characterizing op amp distortion as an input-referred parameter adds great application versatility. The input-referred parameter applies to any application circuit, predicting output distortion through the circuit's $1/\beta$ gain.

Indirectly, the amplifier feedback does have a second-order effect upon the input distortion. Changing the feedback can force changes in signal conditions. Increased closed-loop gain, for example, either increases output signal swing or requires a reduction in input signal, and either signal change alters the dynamic range of signals impressed upon the amplifier's internal nonlinearities. Such nonlinearities may produce higher or lower distortion as the output and input signal ranges change. However, practical applications accommodate the same signal sensitivity in amplifier open-loop gain and common-mode rejection. There the characterization of A_{OL} or CMRR under a given set of signal conditions produces an input-referred parameter that adequately serves any feedback condition, in spite of the actual signal-induced variations. The same convenience holds for distortion since these A_{OL} and CMRR errors also include the distortion signal.

6.2 Basic Distortion Measurement

Op amp distortion characteristics are commonly measured in terms of total harmonic distortion (THD) or intermodulation distortion (IMD).[2] These two measures of distortion face the same equipment and circuit constraints, and the more prevalent THD serves to illustrate these constraints here. THD is defined as 100% times the rms sum of all distortion harmonics divided by the rms value of the signal fundamental.[3] In equation form,

$$\text{THD} = \frac{\sqrt{E_2^2 + E_3^2 + E_4^2 + \cdots + E_n^2}}{E_1} \, 100\%$$

Here E_1 represents the fundamental and E_2 through E_n represent the harmonics. In each case, the capital letter E represents the rms value of the signal, rather than the instantaneous value previously represented by e.

Depending upon the information required, a spectrum analyzer or a distortion analyzer performs the THD measurement. Spectrum analyzers offer detailed distortion information and noise immunity but require longer measurement times. Distortion analyzers permit high-speed measurement but include noise in the measurement and preclude information about the relative effects of individual harmonics.

6.2.1 Spectrum analyzer measurement

Straightforward THD measurements connect a sinusoidal generator E_i at the amplifier input and a signal analyzer at the amplifier output. Figure 6.5 illustrates this connection for the spectrum analyzer case. There the generator delivers the signal E_i to exercise the ampli-

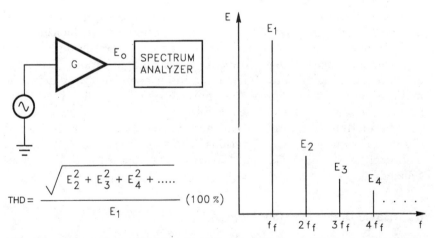

Figure 6.5 With a spectrum analyzer, distortion measurement displays individual harmonics for direct substitution into the THD equation.

fier's distortion sources, and the distortion analyzer processes the resulting amplifier output signal E_o. Distortion products of the amplifier appear in the output signal E_o at the harmonics of the test signal fundamental frequency.

The spectrum analyzer separates E_o into a series of component sine waves and presents these components as a series of amplitude peaks, as shown. The peaks occur at the fundamental signal frequency f_f and at the harmonics of this frequency. Amplitude E_1 at the fundamental frequency represents the remaining pure signal, whereas the other amplitudes represent the distortion harmonics. As a result, the spectrum analyzer measurement produces values for E_1 through E_n that directly fit the previous THD equation. Knowledge of the individual harmonic levels also aids in the evaluation of an amplifier for specific applications. In audio applications, for example, odd harmonics produce a more unpleasant effect than do even harmonics. Also, the spectrum analyzer result virtually eliminates noise from the distortion measurement. The spectral nature of the analyzer measurement restricts noise to a very narrow bandwidth because only the noise in a narrow range about a given harmonic's frequency affects that harmonic's measurement. This feature generally makes noise a negligible effect in the measurement.

6.2.2 Distortion analyzer measurement

Alternately, a distortion analyzer makes the THD measurement for faster results at the expense of noise sensitivity. Figure 6.6 shows the corresponding measurement configuration, and there the analyzer first isolates the distortion signal from E_o selectively. Within the analyzer, a notch filter first removes the signal at the fundamental frequency, leaving the combined signals of the harmonics in the distortion signal E_d. However, signal E_d also includes the broad-band noise of the amplifier, reducing the measurement resolution. To remove some of this noise, the distortion analyzer typically limits the measurement bandwidth. However, the measurement must retain a bandwidth covering the frequency range of all harmonics of interest. Fortunately, op amp noise does not often limit this measurement with a distortion analyzer. Only very-low-distortion op amps or very-wideband measurements encounter an amplifier noise limit. Higher amplifier gain does increase the amplifier noise presented to the analyzer but does not compromise the measurement resolution. Due to the input-referred natures of op amp distortion and noise, the higher gain treats amplifier distortion and noise alike, leaving the ratio of distortion to noise unchanged.

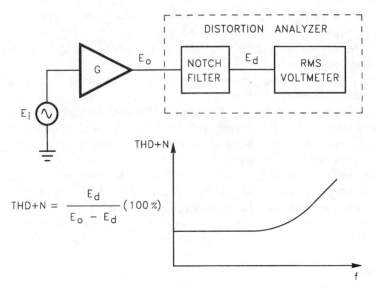

Figure 6.6 With a distortion analyzer, measurement automatically combines harmonics in a distortion signal E_d that also includes broad-band noise.

Following its notch filter, the distortion analyzer simply measures E_d with an rms voltmeter. There the signal measured is

$$\sqrt{E_2^2 + E_3^2 + E_4^2 + \cdots + E_n^2 + E_N^2}$$

where E_N represents the noise signal of the measurement. This measurement reflects THD plus noise, or THD + N, and this combination produces a fortuitous indicator for audio applications. There broadband noise and distortion simultaneously introduce resolution limits, and the detection of either limit serves the purpose.

The distortion analyzer next compares the measured E_d with the fundamental. The fundamental signal consists of E_o less the distortion and noise signals. Paralleling the definition of THD, this measured THD + N is

$$\text{THD} + N = \frac{E_d}{E_o - E_d}\ 100\%$$

Repeating the THD + N measurement at multiple test frequencies produces a curve of distortion versus frequency like the one shown. The flat portion of the curve reflects the effects of dc distortion sources and amplifier noise. At higher frequencies, this curve typically rises with a 20-dB per decade slope, resulting from the decreasing amplifier loop gain and the increasing effects of nonlinear capacitances within the amplifier.

6.2.3 Basic measurement limitations

In practice, limitations of test equipment such as that illustrated in Figs. 6.5 and 6.6 restrict op amp distortion measurement. Distortion introduced by the signal generator and the dynamic range limit of the signal analyzer both limit the measurement resolution. At the start, the signal generator produces distortion products of its own, which represent a certain percentage of E_i. Amplification by the amplifier under test leaves this percentage unchanged, but the generator distortion can easily override the distortion signal to be measured. Meaningful measurement of amplifier distortion with these basic methods requires a signal generator having a significantly lower distortion level. Signal analyzers also impose a measurement limit for these straightforward measurement methods. There the analyzer's finite dynamic range capability limits the resolution of small distortion signals in the presence of the larger test signal. The analyzer must process the large amplitude of the E_1 fundamental and still resolve the smaller amplitudes of the harmonic components. To resolve a harmonic of amplitude E_x, the signal analyzer must have a dynamic range much greater than $(E_1/E_x):1$.

6.3 Feedback Separation of Distortion Signals

Op amp feedback overcomes the limitations of basic distortion measurement through an automatic separation of the distortion and test signals. Section 6.1 demonstrated that all op amp distortion products consolidate in the error signal e_{id}. There the distortion signal appears as an input-referred error, separated from the test signal. This separation removes the generator distortion from the measurement and dramatically reduces the dynamic range of the signal measured. Thus the measurement of e_{id}, instead of e_o, largely avoids the measurement limits imposed by the test equipment.

Analysis quantifies the degree of signal separation achieved. Feedback separates the distortion and test signals to the degree permitted by other amplifier errors. Signal e_{id} contains linear, as well as distortion, error components, and the linear components continue to couple indirectly some degree of generator distortion to e_{id}. Similarly, the linear components continue a background signal against which the distortion components must be measured. Residual limitations depend both upon the op amp's A_{OL} and CMRR and upon the circuit's feedback factor ß. Examination of the follower, noninverting, and inverting op amp configurations produces a universal expression for the reduction in dynamic range requirements and related expressions for the reduction in signal generator requirements for the noninverting and inverting configurations.

6.3.1 Signal separation and the voltage follower

The voltage follower offers the most direct indication of the signal separation benefit. Further, this connection's unity gain permits the maximum signal excitation of both amplifier distortion sources. The two sources, A_{OL} and CMRR, both experience full dynamic range excursions upon application of a full-scale input signal to the voltage follower.

Analysis of Fig. 6.7 quantifies the signal separation benefit for this configuration. This benefit equals the degree of test signal reduction between e_o and e_{id} and for the follower shown,

$$e_{id} = e_i - e_o$$

The follower makes $e_o \approx e_i$ and the preceding subtraction makes $e_{id} \ll e_o$, indicating a significant reduction in measurement dynamic range requirements. As described before, both e_{id} and e_o contain the same amplifier distortion products, and measuring these products in the background of e_{id} places far less demand upon the dynamic range. This subtraction also removes any distortion common to e_o and e_i from the separated signal e_{id}. From a circuit perspective, the follower transfers the generator's distortion in e_i to e_o with very nearly unity gain. Thus the subtraction virtually removes the generator distortion from e_{id}.

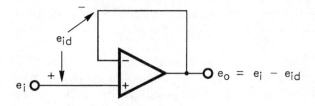

$$e_{id} = e_i - e_o$$
$$= e_o(1/A_{OL} + 1/CMRR)$$

Figure 6.7 Voltage followers separate amplifier distortion from test signals through signal e_{id}, removing generator distortion error and reducing measurement dynamic range.

The approximation $e_o \approx e_i$ also permits a more qualitative evaluation of the signal separation benefit. Examination of the approximation quantifies the benefit in terms of A_{OL} and CMRR. Both amplifier characteristics produce residual feedback signals between the op amp's inputs. Considered separately, the gain and common-mode errors define the op amp error signal as $e_{id} = e_o/A_{OL} + e_{icm}/\text{CMRR}$. For the voltage follower, the common-mode voltage is simply $e_{icm} = e_i \approx e_o$, making the common-mode error term e_o/CMRR. Then the error signal e_{id} becomes

$$e_{id} = e_o(1/A_{OL} + 1/\text{CMRR})$$

Here the e_{id} expression combines the gain and common-mode errors as additive terms, and this represents the worst-case condition. In practice, the common-mode error can be either positive or negative and sometimes subtracts from, rather than adds to, the gain error.

Both A_{OL} and CMRR remain large over most of the amplifier's useful frequency range. There $e_{id} \ll e_o$, reducing the distortion measurement demand on test equipment by a factor of $e_o/e_{id} = 1/(1/A_{OL} + 1/\text{CMRR})$. The signal reduction from e_o to e_{id} directly reduces the dynamic range requirement of the analyzer. Indirectly, the signal reduction attenuates the effect of the signal generator distortion by subtraction, as described earlier. In this way, a typical op amp reduces the test equipment demands by at least 100:1 from dc to around 10 kHz.

6.3.2 Signal separation and noninverting configurations

In the preceding section, the signal separation reduces test equipment demands for the follower by a factor of $1/(1/A_{OL} + 1/\text{CMRR})$. Other op amp configurations add feedback networks that replace A_{OL} here with βA_{OL}, the loop gain. In the noninverting amplifier of Fig. 6.8, the voltage divider action of the feedback network presents a signal $e_o R_1/(R_1 + R_2)$ to the amplifier's inverting input. For the earlier voltage follower case this signal was the full e_o. Now the feedback signal is attenuated, and a simple loop equation shows that

$$e_{id} = e_i - \frac{e_o R_1}{R_1 + R_2}$$

Examination of this expression reveals measurement benefits much like those described for the voltage follower. Signal e_{id} spans far less dynamic range than e_o, and the distortion included in e_i cancels in e_{id} through a combination of forward gain and feedback attenuation. In this combination, the circuit first amplifies the signal e_i and its distor-

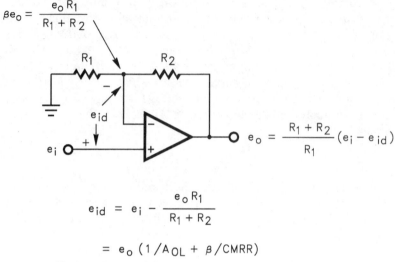

Figure 6.8 Noninverting amplifiers repeat the signal separation of the follower with only a gain change for difference signal $e_i - e_{id}$.

tion, producing $e_o \approx (R_1 + R_2)e_i/R_1$. Then feedback attenuates e_o by the factor $R_1/(R_1 + R_2)$, the inverse of the forward gain. Thus the effects of the forward gain and the feedback attenuation cancel, making the net feedback signal $e_o R_1/(R_1 + R_2) \approx e_i$. This makes $e_{id} \ll e_o$ in the e_{id} subtraction for greatly reduced dynamic range. Similarly, with $e_o R_1/(R_1 + R_2) \approx e_i$, both terms of the equation contain the same generator distortion signal, and the subtraction of this equation again removes the generator distortion.

Examination of the sources of e_{id} quantifies the distortion measurement improvement available with the noninverting amplifier. This differential input signal develops in support of the amplifier's gain and common-mode error signals, making $e_{id} = e_o/A_{OL} + e_{icm}/\text{CMRR}$. For the noninverting amplifier, $e_{icm} = e_i$ and feedback produces an approximately equal signal, $ße_o \approx e_{icm}$, at the op amp's inverting input. Substituting $ße_o$ for e_{icm} reduces the e_{id} equation to

$$e_{id} = e_o(1/A_{OL} + ß/\text{CMRR})$$

Manipulation of this equation reveals the reductions in measurement dynamic range and in sensitivity to generator distortion. Two factors combine to produce these reductions. First the measurement of e_{id}, instead of e_o, reduces the measurement background signal by a factor of (e_o/e_{id}):1. As with the voltage follower, this directly reduces the maximum signal of the measurement dynamic range to increase resolution. However, for the noninverting amplifier, the relative improvement in measurement resolution also decreases due to a second factor. This configuration amplifies the distortion in e_{id} by $1/\beta$, producing a larger distortion signal in e_o. That amplification eases the direct measurement of distortion in the e_o signal and reduces the relative benefit of the signal separation method by the factor of $1:(1/\beta)$. Combining the two factors produces $(e_o/e_{id})/(1/\beta)$ and defines the net dynamic range improvement of the signal separation measurement as a net factor of

$$\frac{\beta e_o}{e_{id}} = \frac{1}{1/\beta A_{OL} + 1/CMRR}$$

Mathematically, this factor simply equals the loop gain βA_{OL} in parallel combination with CMRR. Note that for $\beta = 1$ this equation becomes the reduction factor described for the preceding voltage follower. As will be seen, this equation serves as a general result, applicable to all configurations.

This same factor reflects the reduction in sensitivity to the signal generator distortion. The ratio e_i/e_{id} expresses a reduction as the factor by which e_i is attenuated before appearing in e_{id}. This attenuation reduces the e_i distortion component along with the rest of that signal. With the noninverting amplifier, the feedback network defines the output signal as $e_o = e_i/\beta$, and combining this expression with the reduction factor e_i/e_{id} again yields the net factor $\beta e_o/e_{id}$. Thus the previous expression for $\beta e_o/e_{id}$ also expresses the reduction factor for measurement sensitivity to signal generator distortion.

6.3.3 Signal separation and inverting configurations

For signal separation analysis, the inverting configuration complicates the relationships between e_i, e_{id}, and e_o. Previously, only e_o drove the feedback network, making the feedback signal easy to define. Figure 6.9 shows the inverting configuration, and there both e_i and e_o drive the R_1/R_2 network. Thus both signals influence e_{id} through the feedback network. To find the net result, superposition analysis considers e_o and e_i separately by exercising the feedback network as a voltage divider driven from first one end and then the other. This analysis defines two voltage divider effects, and combining the two produces the superposition result

$$e_{id} = -\frac{e_i R_2}{R_1 + R_2} - \frac{e_o R_1}{R_1 + R_2}$$

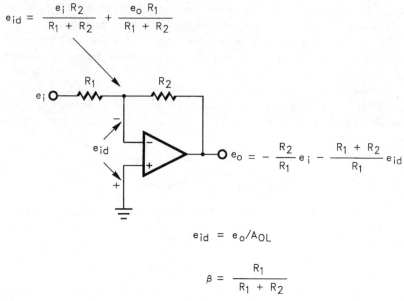

Figure 6.9 Inverting configurations complicate feedback signals but still separate amplifier distortion signal from the test signal through e_{id}.

While it is not immediately obvious from the preceding, the distortion introduced by input signal e_i still cancels in this e_{id} signal. The equation shows that e_i influences e_{id} directly in the first term of the equation, and in the second term, e_i contributes to e_{id} through e_o. There substituting the ideal $e_o \approx -R_2 e_i/R_1$ of the inverting amplifier yields $e_{id} \approx 0$, indicating removal of the e_i effect. This also indicates a reduced signal dynamic range.

More exact analysis quantifies the actual reduction in test equipment requirements, and two familiar factors again combine to define the net reduction. First, the measurement of the smaller e_{id} signal tends to reduce the required dynamic range by a factor of e_o/e_{id}. However, the measurement of e_{id} rather than e_o also reduces the distortion signal measured by a factor equaling the circuit's $1/ß$ gain. This effect tends to increase the dynamic range requirement, counteracting part of the e_o/e_{id} reduction. Dividing e_o/e_{id} by $1/ß$ again defines the net reduction in dynamic range requirements as the factor $ße_o/e_{id}$. Noting that $e_o/e_{id} = A_{OL}$, this factor reduces to the circuit's loop gain,

$$\frac{ße_o}{e_{id}} = ßA_{OL}$$

A related factor expresses the reduction in measurement sensitivity to the signal generator's distortion. The small residual of e_i remaining in e_{id} defines a sensitivity reduction factor of e_i/e_{id}. Setting $e_o = -R_2 e_i/R_1 = (1-1/\beta)e_i$ in the preceding equation reduces it to $e_i/e_{id} = \beta A_{OL}/(1-\beta)$. The added $(1-\beta)$ denominator term of this expression reduces the signal separation benefit of the inverting configuration when compared with the simple βA_{OL} reduction of the earlier noninverting configuration. However, for high closed-loop gains, β becomes small and the two configurations produce similar measurement benefits.

Comparing this result with that of the noninverting case confirms the general equation developed there. The general equation expresses the reduction in test equipment requirements as the factor

$$\frac{\beta e_o}{e_{id}} = \frac{1}{1/\beta A_{OL} + 1/\text{CMRR}}$$

The inverting configuration here makes the common-mode signal zero, removing the CMRR error signal. Removing this CMRR term reduces the general equation to that of the inverting configuration. Similar analysis of any op amp configuration yields either the general equation or a simplified equivalent.

6.4 Direct Measurement of Feedback Error Signal

The preceding section demonstrates the dramatic improvements in op amp distortion measurement achieved through the measurement of e_{id} rather than e_o. The direct measurement of e_{id} offers the greatest relief from test equipment limitations and uses straightforward measurement techniques. Later sections describe alternate measurement methods that simplify the test circuit or extend the measurement resolution further.

Measuring op amp distortion through e_{id} requires mathematical adjustments of the measurement results. The measured result reflects the amplifier's input-referred distortion relative to the residual test signal in e_{id}. However, distortion ratings compare the amplifier's distortion to the full test signal, requiring adjustment of the e_{id} measurement result. Also, the input-referred distortion in e_{id} does not reflect the increased output distortion produced by the circuit's closed-loop gain. Mathematical adjustments correct for these two differences to reflect the amplifier's equivalent output distortion.

For noninverting op amp configurations, a distortion measurement through e_{id} also requires an adjustment to the test circuit. In these configurations, the separated e_{id} signal rides upon a common-mode signal, away from ground. However, the signal analyzer requires a ground-referenced input signal. To accommodate this requirement, the addition of an instrumentation amplifier separates e_{id} from the common-mode signal of the noninverting case, producing a ground-referenced e_{id}.

6.4.1 Direct measurement and inverting configurations

Inverting op amp connections experience no common-mode signal swing and offer the simplest introduction to the mathematical adjustments required when measuring distortion through the e_{id} signal. Figure 6.10 illustrates the distortion measurement for this case with the signal analyzer connected directly to the op amp's summing junction. The signal at this junction is simply E_{id} with no common-mode addition. Here E_{id} designates the rms value of e_{id} to reflect the rms nature of the distortion measurement.

Mathematical adjustments transform the distortion percentage measured in E_{id} to the actual percentage present in E_o. The measured THD, or THD_m, differs from the output distortion THD_o because of the signal separation. The signal separation of E_{id} removes most of the E_o fundamental and reduces the distortion signal to its input-referred level. First, consider the effect of reducing

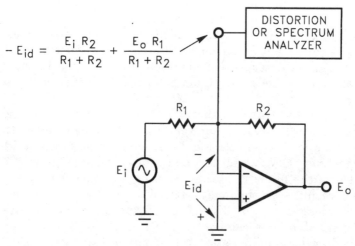

Figure 6.10 For inverting op amp configurations, connecting the signal analyzer to the amplifier's summing junction permits direct measurement of E_{id}.

the fundamental component. The THD calculation compares this component with that of a signal's distortion components. Signal separation reduces the fundamental by a factor of E_{id}/E_o, magnifying the significance of the distortion component by a factor of E_o/E_{id}. Then multiplying the measured result by E_{id}/E_o compensates for the reduced fundamental measured. This adjustment assumes that the signal fundamentals of E_{id} and E_o equal the signals themselves even though both the fundamentals and the distortion components contribute to the signals. Virtually all op amp measurement conditions make this approximation valid. This first adjustment also defines the input-referred distortion as the intermediate result $THD_i = (E_{id}/E_o)THD_m$.

A second mathematical adjustment completes the process to define the output distortion THD_o. This adjustment accounts for the difference between input- and output-referred distortion. Op amp circuits amplify E_{id}, including its input-referred distortion signal, by a gain of $1/ß$. This increases the relative significance of the op amp's distortion signal at the amplifier output. Thus $THD_o = THD_i/ß = (E_{id}/ßE_o)THD_m$.

Implementation of this THD_m to THD_o conversion differs for distortion and spectrum analyzer applications. For distortion analyzer measurements, direct multiplication of the measured result by the factor of $E_{id}/ßE_o$ makes the conversion. Then for $1/ß = (R_1 + R_2)/R_1$,

$$THD + N_o = \left(1 + \frac{R_2}{R_1}\right)\frac{E_{id}}{E_o} THD + N_m = \left(1 + \frac{R_2}{R_1}\right) THD + N_i$$

In this case, the signal level actually measured defines E_{id} and an independent measurement determines E_o. This conversion also corrects for the difference in measured and output noise. Just as with the distortion signal, the signal separation of the measurement magnifies the relative significance of the amplifier noise signal, and the circuit gain amplifies this noise, along with the distortion, in producing E_o. The preceding conversion compensates by supplying the same adjustment to the noise component N and the distortion component THD of THD + N.

For spectrum analyzer measurements, the distortion calculation ignores E_{id} to simplify the analysis. The spectrum analyzer separates the distortion harmonics from the E_{id} fundamental measured. Normally, the distortion calculation includes these various signal components in the equation

$$THD = \frac{\sqrt{E_2^2 + E_3^2 + E_4^2 + \cdots + E_n^2}}{E_1} 100\%$$

Here E_1 represents the fundamental component and would equal E_{id} for the signal actually measured. However, substituting E_o for E_1 instead directly produces the conversion $\text{THD}_i = (E_{id}/E_o)\text{THD}_m$. Then $\text{THD}_o = \text{THD}_i/\beta$, or

$$\text{THD}_o = \frac{(R_1 + R_2)\sqrt{E_2^2 + E_3^2 + E_4^2 + \cdots + E_n^2}}{R_1 E_o} 100\%$$

Note that the signal analyzer connection to the op amp summing junction influences feedback. Connecting the analyzer's input capacitance to this junction can affect both the measurement bandwidth and the circuit's frequency stability. This capacitance loads the feedback network, restricting measurement bandwidth to no more than $f_p = \sqrt{f_c/2\pi R_2 C_i}$. Here f_c is the unity-gain crossover of the op amp and C_i is the net capacitance at the op amp input. The presence of C_i reduces the measurement bandwidth to f_p whenever this frequency is less than the normal amplifier bandwidth of βf_c. Capacitance at the input of an op amp can also degrade frequency stability as described in Sec. 4.2. As described there, capacitive bypass of R_2 restores stability but also reduces bandwidth.

6.4.2 Direct measurement and the voltage follower

The inverting configuration discussed in Sec. 6.4.1 presents E_{id} as a ground-referenced signal, directly accessible by the signal analyzer input. However, noninverting configurations offset E_{id} from ground on a common-mode signal. Then the signal analyzer measurement of E_{id} first requires removal of the common-mode signal.

The voltage follower connection illustrates the common-mode signal, and its removal, in Fig. 6.11. There signal E_{id} rides on the input signal E_i, and this includes E_i in any ground-referenced measurement of E_{id}. Three alternative methods separate E_i from E_{id}, as described in this and the following sections. In this section an instrumentation amplifier simply translates E_{id} to a separate, ground-referenced signal. As shown in Fig. 6.11, the differential inputs of the instrumentation amplifier sense E_{id}, and this amplifier's common-mode rejection removes E_i, translating E_{id} to a ground-referenced signal. Also the instrumentation amplifier amplifies E_{id}, increasing the signal level presented to the analyzer. This signal translation approach achieves virtual immunity to the test equipment limitations described before. The signal separation of E_{id} and the instrumentation amplifier remove E_i and its distortion from the measurement. This signal removal also minimizes the dynamic range of the measurement.

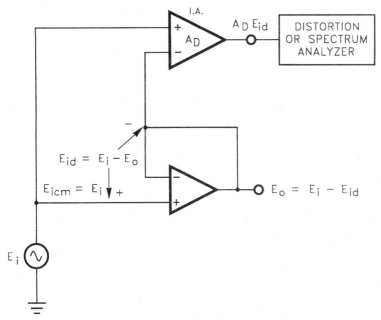

Figure 6.11 With voltage followers, E_{id} rides on a common-mode signal, requiring translation to a ground-referenced signal for distortion measurement.

However, the instrumentation amplifier introduces its own distortion into the measurement, setting a new measurement limit. Common-mode swing at the instrumentation amplifier's inputs reacts with this amplifier's common-mode distortion sources. The full test signal E_i exercises this distortion source, along with that of the amplifier tested. This transfers the distortion measurement challenge from the test equipment to the instrumentation amplifier. As a result, this new measurement limit restricts the technique to the measurement of intermediate levels of op amp distortion. More demanding measurements rely upon the selective amplification and bootstrap alternatives described later.

When the instrumentation amplifier solution applies, two mathematical adjustments of the measured result THD_m translate it to the corresponding output-referred distortion THD_o. These adjustments compensate for the reduced fundamental measured and for the gain of the instrumentation amplifier. First, the measurement of $A_D E_{id}$, rather than E_o, reduces the measured fundamental by a factor of $A_D E_{id}/E_o$. This reduction of the fundamental increases the relative significance of the distortion signal by a factor of $E_o/A_D E_{id}$, and multiplying the measured result by the factor $A_D E_{id}/E_o$ compensates for this effect. Here the A_D term of this factor reflects the amplification of the E_{id} fundamental by the instrumentation amplifier. However, this amplifier also amplifies the distortion components of the E_{id} signal by the same gain, requiring a second adjustment. Multiplying the measured result by $1/A_D$ compensates for the amplified distortion components and reduces the net adjustment multiplier to E_{id}/E_o. Gain A_D drops out of this factor because the amplification leaves the signal's relative distortion unchanged. Then for distortion analyzer results with the voltage follower,

$$\text{THD} + N_o = \frac{E_{id}}{E_o} \text{THD} + N_m$$

For spectrum analyzer results, the calculation of THD includes the adjustments through two changes that adapt the basic THD equation to the E_{id} distortion measurement shown. First, the substitution of E_o for the measured fundamental corrects for the smaller fundamental signal present in the $A_D E_{id}$ signal. This substitution also accounts for the effect of A_D upon the measured fundamental. However, the distortion harmonics measured retain the effect of A_D, and this requires dividing the THD equation by that gain. The two changes to the THD equation adjust the spectrum analyzer results through

$$\text{THD}_o = \frac{\sqrt{E_2^2 + E_3^2 + E_4^2 + \cdots + E_n^2}}{A_D E_o} \, 100\%$$

6.4.3 Direct measurement and noninverting configurations

The voltage follower discussed in Sec. 6.4.2 restricts signal conditions in performance measurements. This configuration inherently makes the common-mode and output signals the same. Switching to the general noninverting configuration adds greater independence in the measurement selection of the two signal levels. This switch does not interfere with the signal separation convenience of E_{id} but does alter the adjustments required for the measured results.

Figure 6.12 illustrates direct op amp distortion measurement with the noninverting configuration, and the circuit shown differs from the follower case only through the action of the feedback network. Both make $E_{icm} = E_i$ and require the instrumentation amplifier to separate E_{id} from this signal. This separation again translates E_{id} to a ground-referenced signal, adapting it to the signal analyzer input. However, the noninverting case shown produces $E_o \approx E_i/\beta = E_{icm}/\beta = (R_1 +$

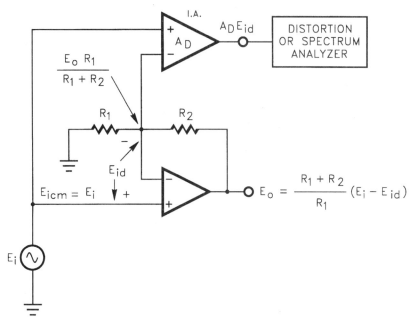

Figure 6.12 Distortion measurement with noninverting amplifiers parallels the voltage follower case, but permits separate selection of common-mode and output voltage swing conditions.

$R_2)E_{icm}/R_1$. Choosing a value for ß sets E_o at any desired multiple of E_{icm}, removing the equal signal requirement of the follower.

As before, mathematical adjustments translate the measured THD_m to the corresponding input- and output-referred equivalents THD_i and THD_o. Two adjustments compensate for the reduced fundamental and distortion components measured. First, the measurement of E_{id}, instead of E_o, reduces the fundamental component of the signal by the factor of E_{id}/E_o, and this increases the significance of the distortion signal relative to the measured fundamental. To compensate, multiplying THD_m by the reduction factor E_{id}/E_o adjusts the result and produces the input-referred distortion of $THD_i = (E_{id}/E_o)THD_m$. A second mathematical adjustment compensates for the reduced distortion signal present in E_{id} as compared with that of E_o. The noninverting configuration amplifies E_{id} by a gain of 1/ß, increasing the actual distortion signal produced in E_o. Multiplying the measured result by 1/ß compensates for this effect, making the net adjustment multiplier E_{id}/E_oß. The gain A_D, provided by the instrumentation amplifier, amplifies both the fundamental and the distortion harmonics, leaving the adjustment multiplier unchanged.

Then $THD_o = (E_{id}/E_o\beta)THD_m$, where $1/\beta = (R_1 + R_2)/R_1$. For a distortion analyzer measurement,

$$THD + N_o = \left(1 + \frac{R_2}{R_1}\right)\frac{E_{id}}{E_o} \, THD + N_m = \left(1 + \frac{R_2}{R_1}\right) THD + N_i$$

For spectrum analyzer measurements, a substitution in the THD calculation replaces the E_{id}/E_o factor of the adjustment multiplier. Here the analyzer indicates the magnitudes of the fundamental and the harmonic components separately. Discarding the indicated E_{id} fundamental and substituting the known E_o replaces the E_{id}/E_o factor. However, discarding the measured E_{id} fundamental also removes the gain A_D of the instrumentation amplifier. The measured harmonics retain this gain, increasing their relative significance. Dividing the modified result by A_D compensates for this gain difference. Then the distortion analyzer equation translates to the spectrum analyzer case as

$$THD_o = \frac{(R_1 + R_2)\sqrt{E_2^2 + E_3^2 + E_4^2 + \cdots + E_n^2}}{R_1 A_D E_o} \, 100\%$$

6.5 Selective Amplification of Feedback Error Signal

Feedback separation of the distortion products expands the distortion resolution through the direct measurement of the E_{id} error signal. For noninverting configurations, however, the direct measurement requires adding an instrumentation amplifier, and the distortion of the added amplifier becomes the new measurement limit. Selective amplification[4] offers an alternative to the added amplifier and removes this new limit. However, this alternative provides less reduction in the measurement demands placed upon the signal generator and the signal analyzer.

The selective amplification approach makes the amplifier tested part of the test system in a way that replaces the function of the previous instrumentation amplifier. With selective amplification, the amplifier tested removes the common-mode voltage, directly providing the required ground-referenced measurement signal. This measurement signal still includes the test signal, but selective amplification increases the relative significance of the distortion signal. In this way, the amplifier tested selectively amplifies the distortion signal to raise this signal above the measurement floor of the test equipment limitations. For inverting circuits, the instrumentation amplifier was not previously required, but these circuits also benefit from selective amplification. In both inverting and noninverting cases, however, the selective amplification reduces the measurement bandwidth.

6.5.1 Selective amplification and the voltage follower

A discussion of the voltage follower connection introduces the selective amplification approach to distortion measurement using Fig. 6.13. There the common-mode rejection of the amplifier tested replaces the instrumentation amplifier used before. This change moves the measurement location back to the op amp's output with its large signal level, but the selective amplification of the op amp's distortion signal eases the measurement. As shown, a bootstrapped feedback modifies the normal follower connection, supplying the selective gain. To do so, resistors R_1 and R_2 form a feedback network that produces gain for E_{id} but not for E_i. Feedback impresses the signal E_{id}, with the amplifier distortion products, on resistor R_1 to produce a feedback current delivered to resistor R_2. This operation develops an error-signal gain of $A_e = 1 + R_2/R_1$ for E_{id} alone. Input signal E_i does not experience this amplification because the bootstrap connection of R_1 isolates this resistor from E_i. In the conventional case, returning R_1 to ground also impresses E_i on this resistor to produce the same amplification as described for E_{id}. However, the bootstrap connection of this case returns R_1 to the output of E_i, removing E_i from R_1 to avoid that added gain. The resulting output signal becomes

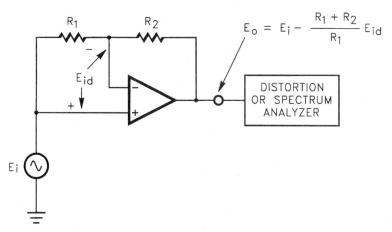

Figure 6.13 Selective amplification of E_{id} magnifies amplifier distortion signals, easing distortion measurement at the amplifier output.

$$E_o = E_i - \left(1 + \frac{R_2}{R_1}\right)E_{id}$$

Due to the bootstrap, the test configuration reflects the test conditions of the voltage follower. Examination of this equation and the preceding figure shows that the amplifier's signal conditions match those of the follower. First, the equation shows that the amplifier output still follows E_i, except for the difference produced by the amplified E_{id} signal. High loop gain minimizes this difference by keeping the error signal E_{id} small. Bandwidth constraints assure the required loop gain, as discussed later. Also, like the follower, signal E_i directly varies the voltage at the amplifier's noninverting input to exercise common-mode rejection distortion sources. To do so, feedback forces the op amp's inverting input in Fig. 6.13 to follow also E_i, and thus the amplifier's input circuit experiences the same common-mode signal as with the basic follower. Combined, the input, output, and common-mode conditions of the test circuit virtually duplicate those of the basic voltage follower.

However, the selective amplification raises the relative magnitude of the E_{id} signal, increasing the resolution of distortion measurement. The preceding equation shows that this selective amplification reduces the requirements for both the signal generator and the signal analyzer. The distortion introduced by the signal generator does remain in the signal measured. However, the selective gain magnifies the distortion of E_{id} by a factor of $(R_1 + R_2)/R_1$, reducing the relative significance of the generator distortion. Similarly, the selective gain reduces the dynamic range required of the signal analyzer. As usual, the full test signal $E_o \approx E_i$ sets the upper extreme of this range, but the selective amplification of E_{id} makes the amplifier's distortion signal a larger part of E_o. This amplification raises the lower extreme of the dynamic range requirement, reducing the overall dynamic range by a factor equal to the selective gain of $(R_1 + R_2)/R_1$.

For this measurement technique, mathematical adjustments compensate the measured result for the effects of the selective gain to define the equivalent output-referred distortion. For the voltage follower considered here, this simply requires dividing the measured THD by the selective gain $(R_1 + R_2)/R_1$. This adjustment converts a distortion analyzer result $THD + N_m$ to the output-referred distortion

$$THD + N_o = \frac{R_1}{R_1 + R_2} THD + N_m = THD + N_i$$

Similarly, for a spectrum analyzer result,

$$\text{THD}_o = \frac{R_1\sqrt{E_2^2 + E_3^2 + E_4^2 + \cdots + E_n^2}}{(R_1 + R_2)E_1} \, 100\% = \text{THD}_i$$

Note that for this voltage follower case, the equal input and output signals make $\text{THD}_o = \text{THD}_i$, where THD_i represents the input-referred distortion of the amplifier.

6.5.2 Bandwidth of selective amplification

At first it would seem that the selective gain should be maximized to achieve the greatest measurement resolution. However, the measurement bandwidth declines as this gain increases due to the associated feedback factor reduction. Section 1.2.3 describes the feedback-factor and bandwidth relationship. With the op amp now part of the measurement system, this amplifier's bandwidth limits the resolution of higher-order distortion harmonics. In compromise, the selective gain should be chosen to be as large as possible within the bandwidth constraint of the measurement.

Making this compromise first requires determining the circuit's feedback factor. Section 1.1 defines the feedback factor ß as the fraction of the amplifier output fed back to the amplifier input. At first, the bootstrap of this case confuses the determination of this fraction since the bootstrap connection of Fig. 6.13 returns R_1 to the generator output rather than to ground. If R_1 were grounded, the fraction of the output fed back to the input would simply be $R_1/(R_1 + R_2)$. With the bootstrap connection, the feedback fraction appears to be unity because the op amp inputs and output all follow E_i. This suggests that the circuit realizes the full unity-gain bandwidth of the op amp.

However, examining the origin of the bandwidth limit demonstrates the reduced bandwidth produced by the selective gain. The amplifier's finite open-loop gain requires feedback support of an error signal E_o/A_{OL} between the amplifier's inputs, and error signal E_{id} includes this gain error. As frequency increases, the open-loop gain A_{OL} rolls off, increasing the E_o/A_{OL} error. This reduces E_o, producing the closed-loop response roll off that defines the circuit's bandwidth. The selective amplification of E_{id} further reduces E_o, as expressed in $E_o = E_i - (R_1 + R_2)E_{id}/R_1$, moving the circuit's bandwidth limit to a lower frequency. The amplification of E_{id} remains the same with or without the bootstrap of R_1, suggesting the same bandwidth for both cases. This, in turn, suggests a feedback factor of less than unity.

Superposition analysis resolves this feedback factor contradiction by separating the effects of input and output signals. Superposition grounding of E_i leaves only E_o to define the signal at the junction of the feedback network. This grounding makes the fraction of the output fed back to the input the same with or without the bootstrap. Then for selective amplification with the voltage follower,

$$\beta = \frac{R_1}{R_1 + R_2}$$

Given this feedback factor, comparing the feedback demand for gain against the available amplifier gain defines the actual bandwidth. Figure 6.14 provides this comparison with overlaid plots of $1/\beta$ and the amplifier's open-loop gain A_{OL}. The $1/\beta$ curve represents the feedback demand for open-loop gain, and when this demand exceeds the available A_{OL}, the amplifier's response declines. At some higher frequency, the amplifier's open-loop roll off drops the available A_{OL} below the $1/\beta$ demand level, rolling off the circuit's response. The intercept of the $1/\beta$ and A_{OL} curves marks the point of equal supply and demand and defines the 3-dB bandwidth for the circuit. There $A_{OL} = 1/\beta$ and the frequency dependence of A_{OL} determines the frequency of this occurrence for the selective gain application here. For most op amps, a single-pole roll off makes $A_{OL} = f_c/f$ in the region of the intercept, where f_c is the unity-gain bandwidth of the amplifier. At the

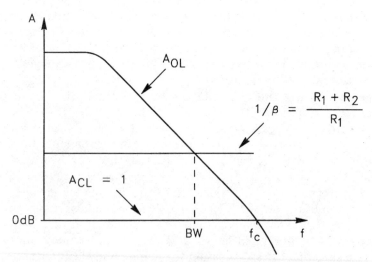

Figure 6.14 Selective amplification in Fig. 6.13 leaves closed-loop gain unchanged but reduces feedback factor and associated measurement bandwidth.

intercept, $f = \text{BW}$, making $A_{\text{OL}} = f_c/\text{BW}$. Combining this result with $A_{\text{OL}} = 1/\beta$ quantifies the circuit bandwidth as

$$\text{BW} = \beta f_c$$

With the bandwidth defined by the measurement requirements, the selection of feedback resistors R_1 and R_2 then optimizes measurement resolution and noise. Here the feedback factor guides the gain selection in the described compromise with bandwidth. Choosing β for the required measurement bandwidth then defines the maximum possible selective gain through $A_e = 1 + R_2/R_1 = 1/\beta$. Typically, the measurement bandwidth must encompass four or five harmonics of the test frequency in order to accurately reflect THD. In this process, the bandwidth requirement defines the ratio of R_2 to R_1, but not the base resistance value. Choosing low base values avoids added noise from these resistors, and these resistors support only small signals, so small resistance values still result in manageable signal currents.

6.5.3 Selective amplification and noninverting configurations

The selective gain approach of distortion measurement extends to the generalized noninverting configuration in Fig. 6.15. There the addition of R_3 produces selective amplification of the E_{id} distortion products. Resistors R_1 and R_2 set the normal closed-loop gain presented to E_i as $A_{\text{CL}} = 1 + R_2/R_1$. Adding R_3 produces greater gain for E_{id} because this signal then develops a feedback current through R_3 as well as through R_1. The parallel combination of these resistors modi-

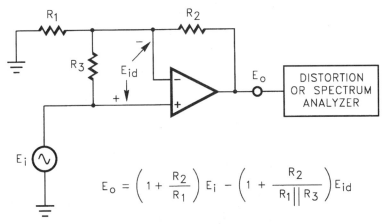

Figure 6.15 Addition of R_3 extends selective amplification of E_{id} to distortion measurements with noninverting configurations.

fies the error-signal gain to $A_e = 1 + R_2/(R_1||R_3)$. As desired, adding R_3 makes $A_e > A_{CL}$ to increase the relative significance of the E_{id} distortion at the amplifier's output. Then the greater output distortion signal again eases the measurement demands placed on the signal generator and the signal analyzer.

In fact, distortion measurement resolution remains unchanged in moving from the preceding follower to the general noninverting configuration here. These circuits differ in the A_{CL} gain supplied to E_i and its distortion, but practical limits equalize the results. For the noninverting case, increasing A_{CL} requires a corresponding decrease in the magnitude of E_i to maintain a given output signal level. This decrease in E_i simultaneously reduces the magnitude of this signal's distortion component by the same factor that A_{CL} increased. The two changes leave the output distortion due to E_i unchanged.

Similarly, the dynamic range requirement for the signal analyzer remains independent of A_{CL}. This range must extend from the level of the fundamental signal down to that of the distortion signal. Independent of A_{CL}, a given test condition defines the output signal level, setting the fundamental. At the other range extreme, the level of the distortion signal measured depends upon the error signal gain A_e. As described before, setting A_e at the maximum level permitted by the bandwidth constraint of the measurement optimizes resolution. This constraint defines A_e and the level of the distortion signal contained in the output. Increasing A_{CL} by reducing R_1 simultaneously requires a reduction in R_3 to maintain a defined level for $A_e = 1 + R_2/(R_1||R_3)$. Thus the lower extreme of the dynamic range also remains independent of A_{CL}.

The noninverting circuit's feedback factor relates A_e to the measurement bandwidth. As described in the preceding section, a feedback factor ß sets the circuit's bandwidth at BW = ßf_c, where f_c is the unity-gain crossover frequency of the op amp. Above this bandwidth limit, higher-order harmonics are attenuated in the measurement. Superposition analysis eases the determination of ß by grounding E_i to determine the fraction of E_o fed back to the amplifier's input. Then for the illustrated noninverting configuration with selective amplification,

$$ß = \frac{R_1||R_3}{R_1||R_3 + R_2}$$

Noting that $A_e = 1 + R_2/(R_1||R_3) = 1/ß$ and combining this with BW = ßf_c yields $A_e = f_c/$BW. Thus for a given f_c, the measurement bandwidth requirement defines the maximum, and optimum, level for A_e.

The selection of R_3 sets this level after first selecting R_1 and R_2 to set A_{CL} and the relative input and output signal swing conditions.

Mathematical adjustments again compensate the measured results for the effect of selective gain. With this gain, the circuit amplifies the input-referred distortion by a gain of A_e, increasing the output-referred distortion measured. In actual operation, the noninverting configuration only amplifies the input-referred distortion by A_{CL}. Multiplying the measured result by A_{CL}/A_e adjusts for this gain difference and for the noninverting configuration

$$THD_o = (A_{CL}/A_e)THD_m = A_{CL}THD_i$$

where $A_{CL} = 1 + R_2/R_1$ and $A_e = 1 + R_2/(R_1||R_3)$.

6.5.4 Selective amplification and inverting configurations

For inverting op amp connections, the preceding noninverting discussion applies almost directly. The two cases differ only in the gain supplied to E_i and in the common-mode voltage delivered to the op amp's inputs. Otherwise the two configurations produce the same distortion measurement results.

Converting the noninverting configuration to inverting simply involves switching the circuit connections to the common return and the input signal. Figure 6.16 illustrates this connection switch with the op amp noninverting input and R_3 returned to ground and with R_1 connected to E_i. As before, resistors R_1 and R_2 set the gain A_{CL} pre-

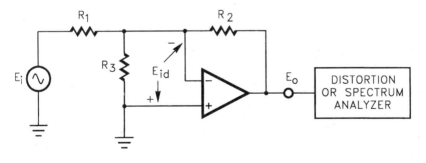

$$E_o = -\frac{R_2}{R_1}E_i - \left(1 + \frac{R_2}{R_1||R_3}\right)E_{id}$$

Figure 6.16 As compared with noninverting cases, selective amplification with inverting configurations alters closed-loop gain and removes common-mode swing.

sented to E_i, but now this gain becomes an inverting $A_{CL} = -(R_2/R_1)$. No change affects the gain presented to E_{id} because the circuit continues to impress this signal upon both R_1 and R_3. The presence of resistor R_3 again boosts the gain for E_{id} to $A_e = 1 + R_2/(R_1||R_3)$.

The higher A_e gain again sets the measurement bandwidth through the corresponding feedback factor ß. Superposition analysis of ß grounds E_i, yielding the feedback factor ß = $(R_1||R_3)/(R_1||R_3 + R_2)$ = $1/A_e$. Then the measurement bandwidth reduces to BW = ßf_c = f_c/A_e, where f_c is the unity-gain crossover frequency of the op amp. This bandwidth limit defines a maximum value for A_e and the resolution improvement provided by this gain. Both the gain and the resolution remain independent of the closed-loop gain A_{CL}, as described in the preceding discussion of the noninverting case.

The higher A_e gain also requires mathematical adjustment of the measured result to reflect the actual output-referred distortion. Gain A_e boosts the distortion signal for easier measurement in the output signal and consequently requires this adjustment to reflect actual performance conditions. Normally, the inverting configuration amplifies the input-referred distortion by a gain of only 1/ß = 1 + R_2/R_1. However, the measurement circuit amplifies this error by the gain A_e. Multiplying the measured result by 1/ßA_e compensates for the difference in measurement and the actual application conditions. For the inverting configuration with selective amplification,

$$\text{THD}_o = \text{THD}_m/\text{ß}A_e = \text{THD}_i/\text{ß}$$

where ß = $R_1/(R_1 + R_2)$ and $A_e = 1 + R_2/(R_1||R_3)$.

6.6 Selective Amplification Alternatives

Two extensions to the selective amplification technique extend the measurement resolution further. First, combining the previous signal separation and selective amplification techniques increases resolution for the inverting configuration. However, this combination does reduce the measurement bandwidth and increases the measurement's sensitivity to test equipment loading. Next, a variable selective amplification approach removes a compromise of bandwidth versus resolution compromise for very-low-distortion cases. There the high selective gain required for resolution of low-frequency distortion restricts the bandwidth for the measurement of higher frequencies. Simply varying the selective gain used at different frequencies resolves the conflict and matches the measurement requirements of practical amplifiers.

6.6.1 Combined signal separation and selective amplification

For inverting op amp configurations, the absence of a common-mode signal permits a combined application of selective amplification and the earlier signal separation technique. In Sec. 6.4, direct measurement of the feedback error separates the amplifier's distortion signal from the test signal to improve measurement resolution. In the preceding section, the selective amplification approach supplies added gain to this error signal to improve also the measurement resolution. Combining the two techniques extends the distortion measurement resolution to the highest level. However, in practice, this combination only serves inverting op amp configurations. In theory, the combined techniques applied here would also serve noninverting configurations. However, the inherent common-mode signal of the noninverting configuration requires the addition of an instrumentation amplifier, which defeats the purpose of very-low-distortion measurement.

For the inverting configuration, combining the two techniques produces Fig. 6.17. There the measurement point returns to the op amp

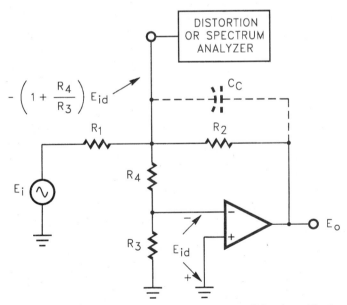

Figure 6.17 Addition of R_4 combines selective amplification with signal separation for inverting configurations.

summing junction, restoring complete signal separation. Also, this measurement circuit includes R_3 to develop a feedback current from E_{id}, just as in the previous selective amplification circuits. However, this circuit does not rely upon R_2 to convert this feedback current to an amplified output error. That would require making the measurement at the amplifier's output in the presence of a large test signal. Instead, the circuit adds an additional resistor, R_4, at the amplifier's input. The feedback current produced in R_3 flows through R_4, producing the desired amplification right at the circuit's summing junction. This connection produces the amplified signal $-(1 + R_4/R_3)E_{id}$ at the top of R_4, and this signal remains separated from the large test signal at the amplifier's output. As before, the measurement of this smaller, separated error signal excludes the signal generator's distortion and reduces the dynamic range required from the signal analyzer.

The distortion measured with this circuit requires two adjustments for conversion to the output-referred equivalent. The first adjustment compensates for the difference in measured and actual fundamental signals as in Sec. 6.4. Multiplying the measured result by $(1 + R_4/R_3)E_{id}/E_o$ adjusts for the difference between the measured $(1 + R_4/R_3)E_{id}$ and the output fundamental E_o. The second adjustment compensates for the gain difference affecting the measured and output signals. The signal measured does not experience the circuit's closed-loop gain of $A_{CL} = (1 + R_2/R_1)$, and multiplying the measured result by A_{CL} compensates for this difference. The two adjustments produce the output-referred distortion expressions

$$\text{THD} + N_o = \frac{R_3(R_1 + R_2)E_{id}}{R_1(R_3 + R_4)E_o} \text{THD} + N_m = \left(1 + \frac{R_2}{R_1}\right) \text{THD} + N_i$$

for distortion analyzer measurements and

$$\text{THD}_o = \frac{R_3(R_1 + R_2)\sqrt{E_2^2 + E_3^2 + E_4^2 + \cdots + E_n^2}}{R_1(R_3 + R_4)E_o} 100\%$$

for spectrum analyzer measurements. Note that E_{id} drops out of this last expression, having canceled with the original denominator voltage term.

Adding selective amplification in this way introduces an added attenuation to the circuit's feedback factor, reducing the measurement bandwidth. In addition to the normal feedback attenuation of R_1 and R_2, a second feedback attenuation results from R_3 and R_4. The latter resistors also produce a loading effect on the attenuation of R_1 and R_2, and the combined effects of the resistor dividers produce a net feedback factor of

$$\text{ß} = \frac{R_1 R_3}{R_1(R_2 + R_3 + R_4) + R_2(R_3 + R_4)}$$

Note that the nonzero values for R_4 that provide selective amplification also reduce ß. This reduces the bandwidth as reflected in Sec. 6.5.2 in the relationship BW = ßf_c, and this limit applies directly to the distortion measurement.

The input capacitance of the signal analyzer C_i also alters this feedback factor. This capacitance bypasses the $R_1 || (R_3 + R_4)$ combination and can cause gain peaking or even oscillation, as described in Sec. 1.4.1. Such problems result if the break frequency of the bypass, $1/2\pi R_1 || (R_3 + R_4)C_i$, occurs within the amplifier's closed-loop bandwidth ßf_c. Adding a compensating capacitor C_C in parallel with R_2 rolls off the gain peaking. However, this compensation capacitor also rolls off the signal bandwidth, limiting the distortion measurement. Choosing this capacitor with the relationship $C_C = R_1 || (R_3 + R_4)C_i/R_2$ optimizes the result by placing the response break of C_C and R_2 at the same frequency that C_i breaks with $R_1 || (R_3 + R_4)$. Then the net feedback divider action remains approximately constant with frequency.

6.6.2 Variable selective amplification

Even with the preceding measurement techniques, some op amp distortion measurement requirements still exceed the test equipment capabilities. This occurs in the measurement of extremely low distortion and in measurements over wider bandwidths. In both cases, the compromise of gain versus bandwidth of the amplifier prevents full characterization of distortion versus frequency with one test configuration. Then variable test configurations produce different gain-bandwidth combinations that match the distortion performance of practical amplifiers.

The selective amplification technique best serves this variable approach. Very low distortion levels automatically rule out the basic signal separation measurement of Sec. 6.4. In that case, the measurement circuits require adding an instrumentation amplifier having an even lower distortion, and that becomes impractical for the measurement of very-low-distortion op amps. Selective amplification avoids the added amplifier but places the measurement bandwidth and measurement resolution in competition. For a given application, the end use defines the bandwidth required for the characterization of distortion versus frequency. Typically, distortion harmonics drop in magnitude at frequencies further away from the test fundamental, making them less significant and limiting the measurement bandwidth requirement. Usually a bandwidth encompassing four harmonics of a given test frequency includes all significant distortion components. Then to use a fixed measurement configuration that covers the entire frequency range of an application, the measurement bandwidth of that configuration must be four times that of the application. For example, the measurement bandwidth must be maintained to around 80 kHz to resolve the harmonics important to the 20-kHz audio range. This bandwidth limits the selective amplification to a gain of $1/ß = f_c/\text{BW} = f_c/80$ kHz, and at lower measurement frequencies, this gain may be insufficient to resolve the lower distortion levels produced there.

However, only the higher-frequency distortion measurements of the characterization require the full bandwidth, and there higher distortion levels ease the requirement for high measurement gain. At lower frequencies, a more restricted bandwidth encompasses the required four harmonics of the test frequency, and this permits higher selective gain levels to better resolve the lower distortion levels encountered there. Measurements at higher test frequencies require the full bandwidth but less resolution. Amplifier distortion typically rises at higher frequencies, reducing the resolution requirement there, and reduced selective gain extends the measurement bandwidth.

Figure 6.18 illustrates the typical measurement requirement and the gain steps that accommodate it. There the THD_i versus frequency plot guides the selection of the circuit's gain steps. As the plot shows, op amp distortion typically rises at the higher frequencies, which require the greatest measurement bandwidth. This distortion rise results from the higher frequency roll offs of the amplifier's A_{OL} and CMRR. These roll offs increase the associated input error signals, increasing THD_i and reducing the need for high measurement gain. In the example shown, varying the selective gain in three steps provides different gain-bandwidth combinations to match the requirements set by the THD_i curve. There the selective gain set by R_1 and R_2 permits distortion measurements with the voltage follower configuration, and varying R_1 changes the gain to accommodate different distortion levels at different frequencies. Analogous measurement conditions for the earlier noninverting and inverting configurations result from varying the resistor R_3.

At lower frequencies, the distortion curve shown remains at its floor level. There distortion measurements require high selective gain but only limited bandwidth. Setting R_1 to R_{1A} establishes this high gain and defines a measurement bandwidth limit $f_1 = ßf_c$, where $ß = R_{1A}/(R_{1A} + R_2)$. Setting this gain at the minimum required for resolution maximizes f_1, limiting the number of gain steps required for the total response characterization. Beyond f_1, the example shown

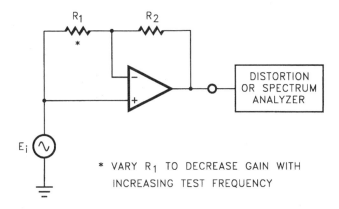

* VARY R_1 TO DECREASE GAIN WITH INCREASING TEST FREQUENCY

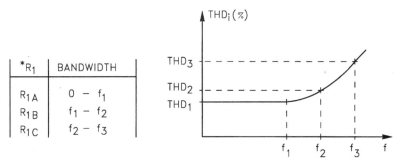

Figure 6.18 Varying the selective gain for different test frequencies adapts the amplifier gain versus bandwidth compromise to varying measurement requirements of the THD_i curve.

decreases the selective gain to increase the bandwidth for the measurement of even higher frequency harmonics. There the rising amplifier THD_i reduces the resolution requirement, permitting this gain decrease. Setting the second gain level at the minimum required for resolution again maximizes the frequency range covered by this setting. Changing R_1 to R_{1B} establishes this second gain level and defines a measurement bandwidth from f_1 to f_2. In this case, the measurement resolution requirement sets the f_1 bandwidth limit and f_2 follows from the normal $BW = \beta f_c$ limit. For even higher frequency tests, changing R_1 to R_{1C} sets a third measurement bandwidth range from f_2 to f_3.

6.7 Bootstrap Isolation of the Feedback Error Signal

The signal separation of op amp feedback provides a complete solution only for the inverting configuration. This configuration directly presents the input error signal of the op amp at the op amp's summing junction. The noninverting configuration also separates the error signal but leaves that signal riding upon a common-mode signal. In Sec. 6.4, an added instrumentation amplifier rejects this common-mode signal but introduces additional distortion. In Sec. 6.5, selective amplification approaches let the amplifier tested remove the common-mode signal, but this approach produces a compromise between resolution and bandwidth, as described in Sec. 6.6.2.

Power-supply bootstrap avoids these earlier limitations for the noninverting configuration. Given care to avoid ground loops, this bootstrap approach eliminates the common-mode signal and extends complete signal separation to the noninverting case. Both the voltage follower and the general noninverting configuration benefit from this approach. Later, combining bootstrap with selective amplification increases the resolution further for the follower configuration.

6.7.1 Bootstrap isolation and the voltage follower

The voltage follower connection provides the simplest introduction to the bootstrap approach to distortion measurement. To demonstrate first the problems resolved by bootstrap, Fig. 6.19 shows the basic measurement connection along with the power supplies, bypass

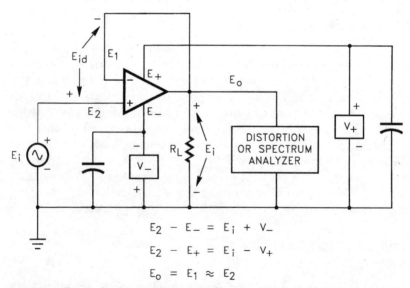

Figure 6.19 Internal voltage swings of an op amp remain relative to power-supply terminals rather than to common.

capacitors, and key node voltages. Including these elements permits the later demonstration of the distortion equivalence of the bootstrap connection. This basic test connection measures the amplifier distortion in the background of the test signal and that signal's distortion. Previously, signal separation by the amplifier feedback removed the test signal for inverting configurations, presenting the amplifier distortion in E_{id} with a much smaller background signal. For that same benefit, the signal analyzer would measure E_{id} directly here. However, the signal analyzer requires a return to common, preventing a direct connection across E_{id}.

To permit this connection, power-supply bootstrap moves the circuit common to the op amp's noninverting input in Fig. 6.20. There the test signal from generator E_i drives what was the common return point before. Otherwise, the voltage follower connection remains unchanged. Conceptually, the common point of a circuit is a relative definition, and this point may be defined anywhere. In practice, the redefinition of common potentially introduces ground loop signals, and the error effects of these signals depend upon the sensitivities of the circuit elements to voltage drops in connecting lines. Here the signal analyzer experiences the greatest sensitivity because it measures

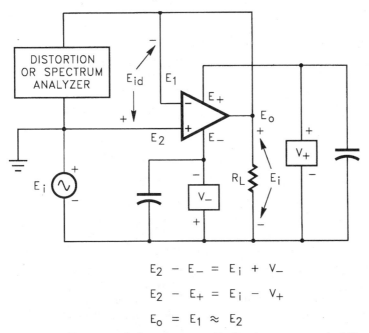

$$E_2 - E_- = E_i + V_-$$
$$E_2 - E_+ = E_i - V_+$$
$$E_o = E_1 \approx E_2$$

Figure 6.20 Power-supply bootstrap permits direct measurement of E_{id} without changing amplifier internal voltage swings.

a small signal in the background of a larger one and, for this reason, the test circuit shown connects the signal analyzer return to the circuit common. In this circuit, the bootstrap connection exposes the power-supply connections to line drops, but the amplifier's power-supply rejection attenuates the resulting effects.

This bootstrap connection again reduces the performance requirements for both the signal generator and the signal analyzer. As before, the distortion measurement moves from the large output signal E_o to the much smaller error signal E_{id}. This move greatly reduces the test signal in the measurement for reduced dynamic range and reduced sensitivity to the generator's distortion. As described in Sec. 6.3.1, this move to the smaller signal reduces the performance requirements for both the generator and the analyzer by a factor of $1/(1/A_{OL} + 1/CMRR)$.

As will be shown, switching to the bootstrap circuit leaves signal conditions, and therefore distortion, unchanged. However, the measurement result from the bootstrap connection again requires an adjustment to account for the reduced fundamental measured. The distortion measurement now detects the fundamental contained in E_{id} rather than that of E_o. Signal E_{id} contains a much smaller signal component at the fundamental frequency, making the measured distortion percentage greater than that of E_o. Multiplying the measured result by E_{id}/E_o compensates for this difference. Note that the bootstrap makes $E_o \approx 0$ relative to common, but this does not represent the signal that exercises distortion sources. Instead, the signal swing E_i across the load resistor represents E_o for this purpose.

The actual implementation of this adjustment depends upon the type of the signal analyzer used. Measurements made with a distortion analyzer directly produce a THD + N_m percentage. Multiplying this percentage by E_{id}/E_i yields

$$\text{THD} + N_o = \frac{E_{id}}{E_i} \text{THD} + N_m = \text{THD} + N_i$$

In this equation, THD + N_o, THD + N_m, and THD + N_i represent the output-referred, measured, and input-referred distortion levels, respectively. A slightly different adjustment compensates for a spectrum analyzer result. Spectrum analyzers individually display the magnitudes of the fundamental and harmonic signals, and normally, substituting these magnitudes into the THD equation yields the circuit's distortion value. However, in this calculation, replacing the measured fundamental with E_i directly compensates for the reduced fundamental actually measured here. This produces

$$\text{THD}_o = \frac{\sqrt{E_2^2 + E_3^2 + E_4^2 + \cdots + E_n^2}}{E_i} \, 100\%$$

6.7.2 Distortion equivalence of the bootstrap measurement

The preceding discussion assumes that the bootstrap connection produces the same amplifier distortion as the basic follower connection. Comparison of the amplifier's signal swings for bootstrapped and nonbootstrapped connections confirms this assumption. The bootstrap connection permits the connection of the signal analyzer directly across E_{id} without changing signal swings. The sensitive analyzer input requires a return to common and, for this, the bootstrap shifts common to the top of the signal generator. Then the generator drives all circuit elements that would normally be returned to common. It drives the returns of the power supplies, their bypass capacitors, and the circuit's load resistor. Otherwise, the bootstrap connection remains the same as the basic follower connection.

Comparison of the signal swings in the two circuits demonstrates their equivalence for distortion measurement. Figure 6.20 reproduces all amplifier signal conditions present in Fig. 6.19. The same input and output voltage swings and the same output signal current exercise the amplifier distortion sources. The bootstrap connection actually holds the amplifier's input and output at or near common with apparently no signal swings. However, the definition of common remains relative, and the op amp itself has no internal return to ground. Instead, the op amp's internal voltages remain relative to the power-supply terminals. Any signal impressed upon or delivered by the amplifier exercises the amplifier's distortion sources through the signal swings relative to the power-supply potentials.

Examining the amplifier's input and output voltages with this new frame of reference demonstrates the signal swing equivalence. Writing loop equations for the bootstrapped case of Fig. 6.20 and the nonbootstrapped case of Fig. 6.19 produces the same expressions for these voltage swings. Both circuits yield an E_2 input signal, relative to the E_- of the negative supply, of

$$E_2 - E_- = E_i + V_-$$

Similarly, both circuits yield an E_2, relative to E_+, of

$$E_2 - E_+ = E_i - V_+$$

Thus the two circuits experience the same internal signals with respect to the E_2 input.

This equivalence extends to the other signal conditions through the inherent operation of an op amp voltage follower. For op amps, the amplifier's two inputs reside at almost the same potential with $E_1 = E_2 - E_{id} \approx E_2$, and this equivalence also applies to input signal E_1. Also the voltage follower connection of the op amp makes $E_o = E_1$ and the preceding equations extend to the output voltage as well. Combined, these op amp and follower conditions also produce the same output signal current for the two circuits. In both cases, loop equations show the voltage across the load to be

$$E_L = -E_{id} + E_i \approx E_i$$

Thus the same input and output signal swings exercise both circuits, and the two produce the same amplifier distortion signal.

6.7.3 Bootstrap isolation and the noninverting configuration

The measurement convenience afforded the follower discussed extends to the general noninverting configuration. As shown in Fig. 6.21, power-supply bootstrap again permits the direct measurement of E_{id} by a grounded signal analyzer. To permit this, the bootstrap connection again places the circuit common at the op amp's noninverting input. In doing so, the bootstrap connection also produces a test signal drive of all circuit elements normally returned to common. All of these bootstrap conditions match those described for the follower before, making the response of noninverting configuration much the same as described for the follower.

However, one circuit difference separates the measurements of the two configurations. The noninverting configuration produces a closed-loop gain A_{CL} greater than unity, and this greater gain produces three measurement differences. First, the greater A_{CL} amplifies E_{id}, making the distortion signal in E_L greater than that measured in E_{id}. Later adjustment of the measured result compensates for this effect. Next, the increased A_{CL} differentiates input and load signals. For the follower, equal input and load signals simplified the demonstration of the distortion measurement equivalence. Here the greater gain produces a larger load signal E_L, requiring reexamination of this equivalence. Finally, the added gain further reduces the performance required of the test equipment. For a given level of E_L, the increased A_{CL} accommodates a reduced level of test signal E_i. The reduced E_i contains a similarly reduced generator distortion effect and produces

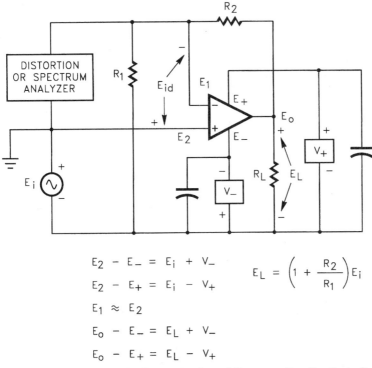

Figure 6.21 Bootstrap of the Fig. 6.20 voltage follower applies directly to distortion measurement of general noninverting circuits.

less E_{id} distortion through signal reaction with amplifier CMRR. As with the preceding voltage follower and other signal separation measurements, the reduced fundamental measured continues to reduce the dynamic range requirement of the signal analyzer. The associated discussion in Sec. 6.3.2 quantifies the distortion effects and demonstrates that the noninverting connection reduces the overall test equipment demands by a factor of $E_L/E_{id} = 1/(1/A_{OL} + ß/\text{CMRR})$.

In this case, the conversion of the measured result to output-referred distortion parallels that described earlier for the follower. However, as described, this conversion requires consideration of the increased A_{CL}. Upon examination, the addition of A_{CL} produces two counteracting effects, leaving the conversion multiplier unchanged. For the follower case before, multiplying the measured result by the factor of E_{id}/E_i compensates for the difference between the measured and the output fundamentals. The noninverting configuration here amplifies the distortion signal in E_{id} by A_{CL}, producing a greater distortion signal in E_o. This effect changes the conversion multiplier to

$A_{CL}E_{id}/E_i$. However, this configuration also amplifies signal E_i by the same gain, making the final conversion multiplier $A_{CL}E_{id}/A_{CL}E_i = E_{id}/E_i$. This multiplier remains unchanged from the preceding follower case, and the previous output-referred THD_o expressions apply here as well. Then for distortion analyzer measurement with any bootstrapped noninverting configuration,

$$\text{THD} + N_o = \frac{E_{id}}{E_i} \text{THD} + N_m$$

or

$$\text{THD}_o = \frac{\sqrt{E_2^2 + E_3^2 + E_4^2 + \cdots + E_n^2}}{E_i} 100\%$$

6.7.4 Distortion equivalence and the noninverting case

The bootstrapped noninverting configuration produces the same amplifier distortion as the nonbootstrapped equivalent. As described with the bootstrapped follower, the op amp distortion products depend upon voltage swings referenced to the power-supply levels. These voltage swings originate in circuit loops not affected by the bootstrap connection. The bootstrap merely moves the point defined as common without changing any circuit loops. The op amp's fundamental return to common remains through the power supplies and not to common directly.

Examination of Fig. 6.21, considering common in two different places, demonstrates the equivalence of the bootstrapped noninverting configuration. With common as shown, loop equations define the amplifier's voltage swings for the bootstrapped case. Then, moving common to the bottom of the signal generator produces the nonbootstrapped case. This change makes no difference in the equations relating the amplifier's voltages relative to the E_+ and E_- supply terminals. Both connections produce the expressions shown for the supply-referenced $E_2 - E_-$ and $E_2 - E_+$ voltage swings. Also, the amplifier's feedback forces $E_1 \approx E_2$, extending these expressions to the second amplifier input. Thus the input-related distortion sources of the amplifier receive the same signal excitation with or without the bootstrap.

Examination of the supply-referenced output signals also demonstrates an equivalent result. There signal E_o relates to E_+ and E_- through the load voltage E_L in a loop formed by R_L, R_1, and R_2. In this loop, feedback develops a signal equal to E_i on resistor R_1. The associated feedback current E_i/R_1 also flows through R_2, producing a voltage of $E_i R_2/R_1$ there. Addition of the voltages on R_1 and R_2 shows the voltage on the load to be $E_L = (1 + R_2/R_1)E_i$. This result portrays the familiar response of a noninverting op amp connection and remains independent of the circuit point defined as common. Moving common to the bottom of the generator does not alter the loop formed by R_L, R_1, and R_2.

Similarly, the loops relating E_o to E_+ and E_- remain unchanged. With common on either side of the signal generator, the output voltage with respect to the two amplifier supply terminals remains

$$E_o - E_- = E_L + V_-$$

or

$$E_o - E_+ = E_L - V_+$$

Thus both the bootstrapped and the nonbootstrapped connections produce the same input and output signal swings for the amplifier. This connection difference leaves distortion products unchanged, and the bootstrapped measurement reflects the distortion performance of the actual application.

6.7.5 Combined bootstrap isolation and selective amplification

For the voltage follower case, a combination of bootstrap and selective amplification achieves even greater distortion resolution. This improvement permits testing the very lowest distortion amplifiers by increasing the level of the signal measured above the analyzer's minimum signal level. Figure 6.22 illustrates the combination with the bootstrap of Fig. 6.20 combined with the selective amplification of Fig. 6.13. Power-supply bootstrap again removes the test signal and selective gain again amplifies E_{id}. The bootstrap connects the amplifier's noninverting input to common, and the basic follower action keeps the amplifier output near zero voltage as well. The near-zero output avoids a reintroduction of the test signal for this selective amplification measurement.

As described with Fig. 6.13, the selective amplification developed by R_1 and R_2 does not alter the gain received by the test signal E_i. In that circuit, E_i transfers from the amplifier input to the follower output with unity gain. With bootstrap, only the amplified error signal appears at the amplifier output and

$$E_o = -(1 + R_2/R_1)E_{id}$$

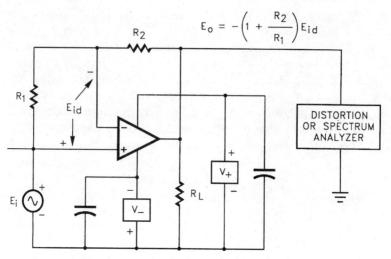

Figure 6.22 Combined bootstrap and selective amplification expand distortion resolution for voltage followers.

Then the signal analyzer measures this amplified error signal, referenced to common, with no interference from the test signal. Further, the amplified error signal overrides the background noise of the analyzer and the measurement environment. This measurement convenience does not extend to the general noninverting case because the added gain there restores a test signal presence in the amplifier's output signal.

Other characteristics of this measurement circuit follow directly from earlier results. As described in Sec. 6.5.2, selective amplification reduces the measurement bandwidth from f_c to $ß f_c$. Here, f_c is the unity-gain bandwidth of the op amp and $ß = R_1/(R_1 + R_2)$ is the circuit's feedback factor. Test equipment requirements remain the same as described earlier for the bootstrapped follower. As before, bootstrap reduces the distortion and dynamic range requirements of the test equipment by a factor of $1/(1/A_{OL} + 1/CMRR)$. No further reduction results from the selective gain because that gain amplifies both the amplifier distortion products and the background signal contained in E_{id}. This signal contains a generator distortion component attenuated by the bootstrap and then amplified by the selective gain along with the amplifier's distortion component. This combination leaves the relative significance of the two components unchanged. Similarly, the selective gain amplifies both the maximum and the minimum signals to be resolved by the analyzer. Thus the dynamic range requirement of the measurement remains unchanged.

For the same reasons, the results measured with Fig. 6.22 translate to output-referred distortion through the same equations presented for the basic bootstrapped follower. The selective gain amplifies both distortion and background signals by the same gain, and the distortion signal remains the same percentage of the total signal before and after this amplification. Thus the earlier results still apply and the output-referred distortion remains for distortion analyzer results,

$$\text{THD} + N_o = \frac{E_{id}}{E_i} \text{THD} + N_m = \text{THD} + N_i$$

and for spectrum analyzer results,

$$\text{THD}_o = \frac{\sqrt{E_2^2 + E_3^2 + E_4^2 + \cdots + E_n^2}}{E_i} \ 100\%$$

References

1. J. Graeme, "Op Amp Distortion Measurement Bypasses Test Equipment Limits," *EDN*, February 17, 1992.
2. D. Pryce, "Audio DACs Push CD Players to Higher Performance," *EDN*, December 7, 1989.
3. The Institute of High Fidelity, Inc., "Standard Methods of Measurement for Audio Amplifiers," IHF-A-202, 1978.
4. J. Graeme, "Advanced Techniques Tackle Advanced Op Amp's Extremely Low Distortion," *EDN*, February 17, 1992.

Glossary

bootstrap Circuit technique that removes the signal voltage swing across a source by driving the source's normal common return with the signal appearing at the source's output terminal.

common-mode input capacitance C_{icm} Effective capacitance between either input of a differential amplifier and common ground.

common-mode rejection (CMR) Logarithmic form of common-mode rejection ratio as expressed by CMR = 20 log(CMRR).

common-mode rejection ratio (CMRR) Ratio of the differential gain of an amplifier to its common-mode gain A_D/A_{CM}.

common-mode voltage Average of the two voltages applied to differential amplifier inputs.

composite amplifier Op amp circuit enclosing two or more op amps within a common feedback loop.

decoupling phase compensation Phase compensation technique that decouples an op amp from a capacitive load through an isolation resistor and a feedback capacitor.

difference amplifier Op amp with a feedback configuration that results in an output signal proportional to the difference of two input signals.

differential input amplifier Amplifier having two inputs of opposite gain polarities with respect to the output.

differential input capacitance C_{id} Effective capacitance between the two inputs of a differential amplifier.

electromagnetic noise coupling Parasitic signal coupling from a magnetic source through mutual inductance.

electrostatic noise coupling Parasitic signal coupling from an electric field through mutual capacitance.

error gain A_e Gain that amplifies the input error signal between an op amp's inputs and equals the application circuit's 1/ß response up to the intercept frequency f_i.

feedback Return of a portion of the output signal of a device to the input of the device for response control.

feedback factor β In a feedback system, the fraction of the output fed back to the input.

feedback intercept *See* intercept frequency.

feedforward factor α In a feedback amplifier, the fraction of an applied input signal fed forward to the amplifier input by the feedback network.

frequency compensation *See* phase compensation.

gain error For an op amp with feedback, the difference between the actual closed-loop gain and that predicted by the ideal gain expression.

gain error signal Differential signal voltage e_o/A developed by feedback between the two inputs of an op amp as a result of the amplifier's finite open-loop gain.

gain margin Difference separating an application circuit's 1/β level and the open-loop gain where the loop phase shift reaches 360°.

input bias current I_B Dc biasing current drawn by both inputs of an op amp.

input capacitance *See* common-mode input capacitance and differential input capacitance.

input error signal e_{id} Combined differential signal voltage developed by feedback between the two inputs of an op amp as a result of the amplifier's error sources.

input offset current I_{OS} Difference between the two input bias currents of an op amp.

input offset voltage V_{OS} Dc voltage impressed between the inputs of an op amp that produces zero output voltage.

instrumentation amplifier Differential input, single-ended output amplifier with internal feedback committed for voltage gain.

intercept frequency f_i Frequency at which an op amp circuit's 1/β response intercepts the amplifier's A_{OL} response, marking the highest frequency for which the amplifier's available gain supplies the feedback gain demand of the application circuit.

loop gain Aβ Excess gain available to supply increasing feedback demand as represented by the separation between an op amp's open-loop response and the application circuit's 1/β response.

minimum-stable gain Minimum gain at which lightly phase compensated op amps can be operated and maintain frequency stability.

multiple feedback Application of multiple feedback connections to form a net feedback factor with greater response control.

mutual capacitance Capacitance that inherently exists between any two objects, where the objects serve as the plates of the parasitic capacitor and the intervening medium serves as the dielectric layer.

mutual inductance Measure of magnetic coupling between two electrical loops in which a generating loop and a receptor loop serve as the primary and secondary of a parasitic transformer action.

noise gain Gain that amplifies the input noise voltage of an op amp and equals the 1/ß response of an application circuit up to the intercept frequency.

offset current See input offset current.

offset voltage *See* input offset voltage.

open-loop gain *A* Ratio of an op amp's output signal magnitude to the signal magnitude appearing between the amplifier's inputs.

phase approximation Straight-line approximation to the phase response introduced by a pole or zero having a slope of ± 45° per decade and crossing the ± 45° point at the frequency of the singularity.

phase compensation Frequency response tailoring for the stability of a feedback system through the addition of response poles and zeros that reduce high-frequency phase shift.

phase margin ϕ_m Margin separating the phase shift around a feedback loop from 360° at the unity loop-gain point of the intercept frequency f_i.

power-supply rejection (PSR) Logarithmic form of power-supply rejection ratio as expressed by PSR = 20 log(PSRR).

power-supply rejection ratio (PSRR) Ratio of a power supply voltage change to the resulting differential input voltage change of an op amp.

rate of closure Difference in slopes of the 1/ß and A_{OL} responses at their crossing as expressed in decibels.

selective amplification Amplification supplied to an op amp's input error signal by virtue of a bootstrapped feedback that does not similarly amplify the applied input signal.

self-resonance Impedance resonance inherent in any passive component due to its parasitic capacitance and inductance combination and especially prevalent in capacitors.

signal separation Input error separation automatically produced by feedback between an op amp's inputs without the background presence of input or output signals.

skin depth Thickness of a given shield material required to attenuate a magnetic field by a factor of $e = 2.73$.

standard denominator 1 + 1/*Aß* Response denominator that transfers the common gain and stability analysis results to all op amp feedback connections.

summing junction Junction of an op amp's feedback elements that drives the amplifier's inverting input and permits signal summation through feedback control of the junction's voltage.

total harmonic distortion (THD) RMS summation of the Fourier harmonic magnitudes of a signal divided by the magnitude of the signal's fundamental times 100%.

transmission block Feedback modeling representation of the α and ß transmission factors of a feedback system.

unity-gain crossover f_c Frequency at which the open-loop gain of an op amp crosses unity.

virtual ground Ground-like characteristic of the feedback input of an op amp at which feedback absorbs injected current without developing a voltage.

voltage follower Short-circuit feedback connection of an op amp which results in an output signal that follows the signal applied at the amplifier's noninverting input.

Index

Analyzer:
  distortion, 179
  dynamic range, 181
  spectrum, 178

Bandwidth, 11
  capacitance load effect, 109, 111
  decoupling phase compensation, 115
  error signal, 12, 56
  feedback-factor compensation, 143
  inverting amplifier, 34
  noise, 36
  noninverting amplifier, 6, 128, 139
  pole-zero compensation, 119
  summing amplifier, 67
  universal limit, 12, 56
  with variable feedback, 64
Bootstrap, 219
  distortion equivalence, 211, 214
  in distortion measurement, 207
  noninverting amplifier, 212
  selective amplification, 215
  voltage follower, 208
Bypass:
  feedback loop, 112
  power supply, 73

Capacitance:
  loading, 108
  mutual, 156
CMRR, 4, 219
Coaxial returns, 168
  ground plane, 170
Common-mode rejection, 219
  of electrostatic coupling, 156

Common-mode rejection (*Cont.*):
  error, 4
  of magnetic coupling, 162, 164
  nondifferential, 158
Composite amplifier, 19, 60, 219
Coupling:
  electrostatic, 156, 219
  magnetic, 160, 219
  RFI, 160
Current source, 50

Decoupling:
  phase compensation, 111, 113, 219
  power supply, 100
Denominator, standard, 9, 55
Difference amplifier, 132, 219
Distortion, 171
  basic measurement, 177
  bootstrap isolation, 207, 215
  common-mode rejection, 176
  direct measurement, 187
  input-referred, 175, 177
  inverting amplifier, 185, 188, 201
  measurement bandwidth, 197
  measurement limits, 181
  nature of, 172
  noninverting amplifier, 173, 183, 192, 199, 208, 212
  open-loop gain, 176
  selective amplification, 194, 202, 203, 215
  signal, 173
  signal generator, 182
  signal separation, 181, 203, 208
  THD+N, 180
  total harmonic, 178, 221

Distortion (*Cont.*):
  voltage follower, 173, 182, 190, 195
Divider, 63

Electrostatic coupling, 156, 219
Electrostatic shielding, 156
Error gain, 6, 56, 219
Error input, 4

Feedback analysis:
  common-mode, 57
  complex, 57
  differential, 57, 58
  generalized, 54
  inverting, 33, 57
  multiple-amplifier, 58
  multiple-input, 65
  noninverting, 8, 57
  process, 48
  summing amplifier, 65
Feedback factor, 2, 32, 220
  combining, 42
  dual-input, 44
  intercept, 11, 220
  multiple-feedback, 67
  net, 50, 67, 70
  op amp influence, 21
  phase compensation, 140
  positive, 22, 38
  signal separation, 181
  summing amplifier, 66
  variable, 63
Feedback model:
  Black, 8
  constructing, 49
  dual-input, 46
  error in, 22, 49
  generalized, 54
  inverting, 30, 32
  multiple-amplifier, 60
  noninverting, 6
  polarities of, 33, 49
  and positive-feedback, 41, 54
Feedforward factor, 33, 56
  summing amplifier, 65
Filter:
  decoupling, 115
  power supply, 100
  PSRR, 94
  voltage-controlled, 68

Frequency response:
  bypass interaction, 96
  power supply noise, 76
  power supply resonance, 83
  supply line impedance, 91

Gain:
  balancing, 134
  closed-loop, 9, 34, 35, 56
  combined-feedback, 43
  error, 13, 35, 56, 62, 220
  loop, 9, 62
  margin, 25, 220
  minimum-stable, 39, 144, 220
  open-loop, 4
  peaking, 126
  selective, 44, 221
Grounding, 156

Inductance:
  mutual, 160, 161, 220
  parasitic 75
Input capacitance, 120, 220
  difference amplifier, 132
  inverting amplifier, 121
  noninverting amplifier, 126
Intercept, 11, 220
  bypass impedance, 97
  dual bypass, 93
Inverting amplifier, 33
  distortion, 185, 188, 201

Line impedance:
  power supply, 73
  stability requirement, 80
Loop area, 162
  matching, 166

Magnetic coupling, 219
  prevention, 167
Magnetic loop, identifying, 163
Magnetic shielding, 161
Multiple feedback, 67, 220
Mutual capacitance, 156, 220
Mutual inductance, 160, 220

Noise:
  gain, 6, 76, 221

Noise (*Cont.*):
   power supply, 74, 76
Noninverting amplifier, 2
   alternate compensation, 136
   distortion, 173, 183, 192, 199, 208, 212

Oscillation:
   intuitive evaluation of, 18
   phase compensation, 107
   requirement for, 13, 18, 79
   supply line coupling, 78
Output impedance, 113

Parasitic feedback loop, 78, 79
Parasitics, capacitor, 87
   inductance, 75, 87, 93
   resistance, 82, 83, 87
Peaking, compensation, 131
Permeability, 161
Phase approximation, 15, 221
Phase compensation, 107, 221
   capacitance load, 108
   decoupling, 111, 113
   difference amplifier, 132, 152
   feedback-factor, 140
   input capacitance, 120
   integrator, 145
   inverting, 125
   multipurpose, 139
   negative feedback, 140, 144
   noninverting alternative, 139
   noninverting configuration, 126, 128
   noninverting test, 131
   pole-zero, 116, 118
   positive feedback, 151
   voltage follower, 148
Phase margin, 110, 221
   capacitance load, 111
Power supply bypass, 73
   design selection, 85, 92
   detuning, 94
   detuning design, 97
   dual, 90
   guideline, 86, 92
   model, 74, 87, 91
   primary selection, 81
   requirement for, 73, 74
   secondary, 90
   secondary selection, 86

Power supply decoupling, 100
   design, 104
   L-C, 102
   R-C, 101
   R-L-C, 103
PSRR, 4, 77

Radiated interference, 155
   trouble shooting, 167
Rate-of-closure, 84, 221
Rejection:
   common-mode, 4, 134, 219
   power-supply, 4, 77, 221
Resonance:
   bypass interaction, 95
   detuning, 94, 97
   dual bypass, 91
   graphical analysis of, 83
   primary bypass, 81
   self, 86
RF detector, 166
RFI coupling, 166

Selective amplification, 194, 221
   alternatives, 202
   bootstrapped, 215
   and measurement bandwidth, 197
   variable, 205
Selective gain, 44, 221
Self resonance, 86
Shielding:
   electrostatic, 156
   magnetic, 161
Signal separation, 208, 221
   inverting amplifier, 185
   noninverting amplifier, 183
   voltage follower, 173, 182, 208
Skin depth, 162, 221
Slew rate limit, 12, 39
   extending, 140, 146
   integrator, 145
Source resistance, 4
Stability, 13, 18
   capacitance load effect, 108
   input capacitance effect, 21, 120
   input capacitance test, 124
   input inductance effect, 24
   power supply coupling effect, 78
   variable feedback effect, 64

Transmission block, 32, 222

Voltage follower, 39, 45, 222
　phase compensation, 116, 148
　　distortion, 173, 182, 190, 195, 208, 215

ABOUT THE AUTHOR

Jerald Graeme is Principal Engineer for the Gain Technology Corporation in Tucson, Arizona. He has 30 years of design, management, and training experience in linear IC product development, and holds eight U.S. patents. Mr. Grame is the author of four bestselling circuit design books for McGraw-Hill, including *Photodiode Amplifiers* and *Designing with Operational Amplifiers*. He has also written more than 100 articles for *EDN* and *Electronic Design*, and was named *EDN's* Innovator of the Year in 1993.